AA002349

Proceedings

The 2006 International Conference
on MEMS, NANO and Smart Systems

December 27 - 29, 2006, Cairo, Egypt

Edited by

Wael Badawy, Dept. of ECE University of Calgary, Canada

Ashraf Salem, Ain Shams University, Egypt

IEEE Catalog Number: 06EX1659
ISBN: 1-4244-0899-7

© 2006 IEEE. Personal use of this material is permitted. However, permission to reprint/republish this material for advertising or promotional purposes or for creating new collective works for resale or redistribution to servers or lists, or to reuse any copyrighted component of this work in other works must be obtained from the IEEE.

Technical Support:

Wael Badawy, Dept. of ECE University of Calgary, Canada; email: badawy@ucalgary.ca
Ashraf Salem, Ain Shams University; email: ashraf@asunet.shams.edu.eg

2006 International Conference on MEMS, NANO, and Smart Systems

Cairo, Egypt
26-28 December 2006

IEEE Catalog Number: CFP06327-POD
ISBN: 978-1-42440-899-3

Table of Contents

Table of Contents	iii
Welcome Message from the Chairs	v
Conference Committee	vii
Keynote Speakers	vii
Industrial Panel	x

Technical Papers

Flow Field Visualization by Micro-PIV in an X-shaped Microfluidic Channel with Two Streams — 1
Jin-Cherng Shyu, and Falin Chen; Industrial Technology Research Institute, Hsinchu County, Taiwan, ROC.

Mechanical Strain Measurements Using Semiconductor Piezoresistive Material — 5
Ahmed A. S. Mohammed, Walied A. Moussa, and, Edmond Lou; University of Alberta, Edmonton, Alberta, Canada.

A New Simple Model for the Single-Electron Transistor (SET) — 7
M. Y. A. Ismail, and R. A. Abdel Rassoul, the Arab Academy for Science and Technology, Alexandria, Egypt.

An Advanced Model for Dopant Diffusion in Polycrystalline silicon during rapid thermal annealing — 11
|S. Abadli and F. Mansour; the department of Electronics, University Mentouri Route d'Ain EL-Bey Constantine, Constantine, Algeria.

Growth of Grains Effect on Boron Diffusion in Heavily Implanted Polycrystalline silicon Thin Films — 15
S. Abadli and F. Mansour; the department of Electronics, University Mentouri Route d'Ain EL-Bey Constantine, Constantine, Algeria.

Analysis and Modeling of RF-MEMS Disk Resonator — 19
Mostafa M. Sakr, Mousa Khalid El-Shafie, and Hany Fikry Ragai, Ain Shams University, Electronics & Comm. Department, Cairo, Egypt.

Behavioral Modeling of RF-MEMS Disk Resonator — 23
Mousa Khalid El-Shafie, Mostafa M. Sakr, and Hany Fikry Ragai, Ain Shams University, Electronics & Comm. Department, Cairo, Egypt.

Determining the Required Pulses for Controlling the Operation of Electrostatic MEMS Converters — 27
Marwa S. Salem*, Mona S. Salem*, A. A. Zekry*, H. F. Ragai; *Ain Shams Univ., Fac. of Eng., ECE Dept., Egypt.

Wafer Level Package Using Polymer Bonding of Thick SU-8 Photoresist — 31

Kyounghwan Na, Ihwan Kim, Eunsung Lee, Hyeon Cheol Kim, Yong-Hwan Lee, Kukjin Chun; School of Electrical Engineering and Computer Science, Seoul National University, Seoul, Korea.

Complete Analysis of a Novel Fully Symmetric Decoupled Micromachined Gyroscope   35
Abdelhameed Sharaf[1,2]; Sherif Sedky[2,3]; S. E.-D. Habib[4,1];
[1]NCRRT, EAEA, Cairo, Egypt; [2]STRC, AUC, Cairo, Egypt; [3]Physics Department, AUC, Cairo, Egypt; [4]Electronics and Communications Dept., Faculty of Engineering, Cairo University, Giza, Egypt.

A New Design of a Current-mode Wheatstone Bridge Using Operational Floating Current   41
Conveyor
Yehya H. Ghallab, Wael Badawy; Biomedical Eng. Dept., Helwan University, Cairo, Egypt, and Department of Electrical and Computer Engineering University of Calgary, Canada.

Performance Enhancement of Gap Closing Electrostatic MEMS Converters   45
Mona S. Salem*, Marwa S. Salem*, A. A. Zekry*, H. F. Ragai*; *Ain Shams Univ., Fac. of Eng., ECE Dept; Egypt.

Highly Efficient Micro-machined Bragg Mirrors Using Advanced DRIE Process   48
B. SAADANY, D. KHALIL, and T. BOUROUINA; ESIEE/Esycom-Lab, Cedex FRANCE; Ain-Shams University, Cairo, Egypt.

Monte Carlo Simulation of Photonic Band Gap Structures   52
Tarek Badreldin and Diaa Khalil, Mentor Graphics, Egypt. Faculty of Engineering, Ain Shams University, 1 El-Sarayat Street, Abbassia, Cairo, Egypt.

Preparation of Polyimide Nanofibers by Electrospinning   58
Mingyan Zhang, Zhaoli Wang, and Yujun Zhang; Harbin University of Science and Technology, Harbin, Heilongjiang Province, People's Republic of China

Welcome Message from the Chairs

The International Conference on MEMS, Nano, and Smart Systems provides a forum for the discussion of new developments, recent progress, and innovations in the design and implementation of MEMS, NANO, and Smart Systems-on-Chip. It addresses all aspects of design methods of those systems. With these goals and applications in mind, Drs. Wael Badawy and Walied Moussa organized the first conference in Banff, Canada in 2006. The success of this first conference led to a yearly International conference organized Banff, Alberta, Canada 2003 – 2005.

This year and due to the many events organized for the opening on the Canadian National Institute for Nana Technologies, the committee decided to more the conference to Cairo, Egypt leaving out the space to these Canadian events and co-locate the conference with the International Workshop on System on Chip. The aim is to keep the conference running as a yearly event.

Cairo, Egypt, the Triumphant City, known officially as al-Qāhirah is one of the world's largest urban areas and offers many sites to see. It is the administrative capital of Egypt and, close by, is almost every Egypt Pyramid, such as the Great Pyramids of Giza on the very edge of the city. But there are also ancient temples, tombs, Christian churches, magnificent Muslim monuments, and of course, the Egyptian Antiquities Museum all either within or nearby the city. Cairo, is an amazing city full of life and movement, and it is that way almost 24 hours every day, with the noisy honking of horns, children playing in the streets and merchants selling their wears and services. And here, the Egyptians are most at home in this powerful, modern and ancient city. It provides great culture, including art galleries and music halls, such as the Cairo Opera House, as well it should, being one of the largest cities in the world. It also provides some of the grandest accommodations and restaurants in the world, such as the Four Seasons and the Cairo Marriott. It also offers an incredible selection of shopping, leisure and nightlife activities. Shopping ranges from the famous Khan el-Khalili souk, (or bazaar) largely unchanged since the 14th century, to modern air-conditioned centers displaying the latest fashions. All the bounty of the East can be here. Particularly good buys are spices, perfumes, gold, silver, carpets, brass and copperware, leatherwork, glass, ceramics and mashrabiya. Try some of the famous street markets, like Wekala al-Balaq, for fabrics, including Egyptian cotton, the Tentmakers Bazaar for appliqué-work, Mohammed Ali Street for musical instruments and, although you probably won't want to buy, the Camel Market makes a fascinating trip. This is, and has been for over a thousand years, truly a shopper's paradise.

The conference received 62 full paper submissions where only 15 papers have been selected for publication after a peer review process with at least two reviews per paper. The conference will be held in conjunction with the International Workshop on System on Chip. We are fortunate to have two distinguished plenary speaker, Managing Director, Mentor Graphics Egypt, Dr. Hazem El Tahawy; and iCORE Chair in Advanced

Technology Processing Systems (ATIPS), Prof. Graham Jullien. Furthermore, the workshop features an industrial panel titled ""Electronics Design Industry in Egypt: Opportunity & Challenge".

Finally we would like to thank the technical co-sponsor of the IEEE Circuits and Systems Society; and the Cooperation of the IEEE Circuits and Systems Society's Technical Committee on VLSI and the IEEE Circuits and Systems Society's Technical Committee on Communication. Also, would like to thank the support of our valuable co-sponsors: Mentor Graphics, University of Calgary, The Technical Institute on Micro, Nano, and Smart Systems, AB, Canada and Ain Shams University

We are delighted to welcome you to this exciting workshop and to Cairo. Have a great time here,enjoy the high quality technical program, and, most of all, enjoy the beauty of the local region.

Wael Badawy, Dept. of ECE University of Calgary, Canada
Ashraf Salem, Ain Shams University, Egypt

December 2006

Conference Committee

Conference Co-chairs

Wael Badawy, Dept. of ECE University of Calgary
Ashraf Salem, Ain Shams University, Egypt

Program Co-chairs

Wael Badawy, Dept. of ECE University of Calgary
Ashraf Salem, Ain Shams University, Egypt

Local Organizing Chair

Ihab Amer; German University in Cairo

Local Arrangement Chair

Mrs. Marwa Zaghow, Mentor Graphics Egypt

Keynote Speakers

Keynote 1

Title:
System on Chip (SOC) design pressures reach critical point

Presenter:
Dr. Hazem ElTahawy; Managing Director; Mentor Graphics Egypt,

Abstract and bio are not available at the printing time.

Keynote 2

Title:

SoC - what are our technology futures?

Prof. Graham Jullien, iCORE Chair in Advanced Technology Processing Systems (ATIPS), University of Calgary

Abstract:

Many of us discovered that we were working in System-on-Chip technology by default! The technology to put hundreds of millions of transistors on a monolithic CMOS digital chip became available over the past few years; we adopted it and became de facto SoC researchers. However, as SoC has come on stream we have also started to see cracks appearing in the technology. 3rd order effects of just a few years ago have become predominant problems, and previous performance predictions have been shown to be false. This talk will undoubtedly produce more questions than answers, but, as an interested observer of the technologies we play with in our sand box, I will try to ponder on some of the issues - and muse on how those amongst us, who normally only observe, can also play a role in defining and exploring promising future technologies.

Brief Bio:

Graham Jullien holds the iCORE Chair in Advanced Technology Information Processing Systems, and is the Director of the ATIPS Laboratories, in the Department of Electrical and Computer Engineering at the University of Calgary. His long-term research interests are in the areas of Integrated Circuits (including SoC), VLSI Signal Processing, Computer Arithmetic, High Performance Parallel Architectures, and Number Theoretic Techniques. Since taking up his chair position at Calgary in 2001, he has expanded his research interests to include security systems, nano-electronic technologies and bio-medical systems. He is currently involved, along with his colleagues, in developing an Integration Laboratory cluster to explore next generation integrated microsystems.

Dr. Jullien is a Fellow of the IEEE, and a member of the Boards of Directors of DALSA Corp., CMC Microsystems and Micronet R&D. He has published more than 380 papers in refereed technical journals and conference proceedings, and has served on the organizing and program committees of many international conferences and workshops over the past 35 years. He was a guest editor of the recent IEEE Proceedings special issue on SoC - Integration and Packaging. Prior to joining the University of Calgary, he was at the University of Windsor, where he held the positions of University Professor in the Department of Electrical and Computer Engineering, and Director of the VLSI Research Group.

Industrial Panel

"Electronics Design Industry in Egypt : Opportunity & Challenge"

Purpose and objectives :
The purpose of this panel is to explore the problems facing this industry in the Egypt and to give some recommendations in the light of the experiences of newly industrialized countries from North East and South East Asia as well as India. The main factors affecting the promotion of this industry in these countries lied in encouraging foreign direct investment, encouraging Research and Development (R&D), improving marketing strategy, enhancing training and human development, and availing sources of finance needed for encouraging capital intensive industries. The problems facing the improvement of the electronics industry in Egypt may be the low levels of know-how, absence of planning for R&D, ineffective current laws in protecting intellectual property rights, inadequate supportive and complementary industries, inwards orientation strategy, and inadequate productive potentiality. Nevertheless, there are some points of strength reflected in low prices of human capital, the strategic place of Egypt, the presence of a reasonable base of infrastructure, and low prices of land and energy in comparison to international prices.

Panel Coordination and Organization
- Dr. Hazem El Tahawy, Mentor Graphics Egypt
- Mrs. Marwa Zaghow, Mentor Graphics Egypt

Panel Frame :
- Coordinator Present the Panel title, objectives and goal as well as Panelist (their affiliation, background and experience) : Mrs Marwa Zaghow
- Coordinator will present the concept behind this panel and the factors to be considered for each presentation : Dr. Hazem El Tahawy
- Each panelist will cover for a presentation of 7 Minutes his experience and his point of view on the Challenge and Opportunity in the Region with some recommendation to improve this industry in the region
- Open Discussion and Questions from the audience

List of Panelist
- Dr. Khaled Elamrawi Intel Corporation (UK) - Middle East, Turkey and Africa
- Dr. Gamal Aly (Ministry of Communication and Information Technology)- Egypt
- Eng. Mohamed Kamal Abdel Fatah (Electronics Factory-Bahgat Group)- Egypt
- Dr. Hisham Haddara, SWS
- Eng, Adel Adib, Alfa Electronics
- Prof.. Ashraf Salem (Ain Shams University)- Egypt

Flow Field Visualization by Micro-PIV in an X-shaped Microfluidic Channel with Two Streams

Jin-Cherng Shyu, and Falin Chen

Abstract—In order to well design a microfluidic fuel cell, a preliminary study of flow field visualization by micro-PIV in an X-shaped microfluidic channel with two streams was performed. Both streams with either the same or different flow rates were fed through two branches of an X-shaped microchannel. The visualization results show that the multi-streams for all tests are kept laminar over entire channel even if the different flow rates are fed. Besides, the X-shaped microchannel was confirmed that such design could separate two streams without any mixing at the exit of the main channel for further utilization of these liquids.

I. INTRODUCTION

With the advance of micro-fabrication technology (in terms of the potential complexity of an integrated device and the ease with which a simple prototype can be made), the development of the microfluidic device becomes inevitably for its inexpensive and versatile features. Typical applications such as DNA analysis [1], cooling [2], micro fuel cell [3] [4], and the like were already demonstrated in the literatures.

Among these aforementioned applications, the development of micro fuel cells is of particular interest for offering a cleaner, more-efficient alternative to the conventional combustion of gasoline and other fossil fuels. Furthermore, it can provide power in stationary and portable power applications. However, miniature of the traditional PEMFCs may suffer from the proton exchange membrane, in which water management and fuel crossover through the membrane [5] could lead to a catastrophe of the whole system. As a result, resolving these problems are still the main concerns of PEMFCs applications, especially those portable applications.

Therefore, a novel microfluidic fuel cell has thus been proposed to eliminate the membrane via transporting the fuel and oxidant streams in a single micro-channel under laminar flow condition. A schematic of this concept can be seen in Fig. 1, the two streams are separated by a liquid–liquid interface [5] ~ [10]. Apparently such design has certain advantages over static membrane fuel cells since convective transport dominates over diffusive transport during operation, so fuel crossover can be avoided. In addition, water management is not needed due to the automatic removal of the excess water generated in the electrochemical process by the flowing streams. Furthermore operation at elevated temperatures is not a problem provided that the boiling point of the chosen fuel is applicable. Another phenomenon occurring is the formation of depletion boundary layers close to the catalyst-covered electrodes as a result of the reactions of both fuel and oxidant at the corresponding electrodes. Adjustments of flow rates and channel dimensions allow precise control of the electrochemical processes that is taken place at the electrodes.

By a careful examination of the recent researches on such membraneless micro fuel cell, one would observe that a micro-channel with 0.5 ~ 1.0 mm in both width and depth is usually fabricated with the electrodes being placed on either the undersurface or the sidewalls of the micro-channel to form a microfluidic fuel cell system. Besides, the flow rate of both fuel and oxidant streams is always kept the same ranged from 0.1 ~ 2 mL/min.

Note that both the micro-channel design and flow rates are important design parameters affecting the performance of such microfluidic fuel cells, and the separation of fuel and oxidant streams after leaving the micro-channel must be recycling for efficient fuel utilization. Hence, it is crucial to investigate the convective transport phenomenon of the two parallel streams with either the same or different flow rates in an X-shaped micro-channel. It is the objective of this study to perform a flow visualization experiment within this microfluidic device for designers' reference.

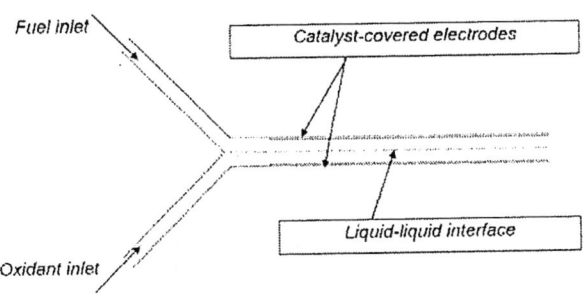

Fig. 1 Schematic diagram of a microfluidic fuel cell

Manuscript received August 1, 2006.

Jin-Cherng Shyu is a researcher with the Energy and Environment Research Laboratories, Industrial Technology Research Institute, Hsinchu County, Taiwan 31040, ROC (phone: +886-3-5918549; fax: +886-3-5829782; e-mail: jcshyu@itri.org.tw).

Falin Chen is now the vice president and general director with the Energy and Environment Research Laboratories, Industrial Technology Research Institute, Hsinchu County, Taiwan 31040, ROC (e-mail: falinchen@itri.org.tw).

II. EXPERIMENTAL SETUP

In order to perform the present flow field visualization, an experimental apparatus composed of three major systems were established, including a microfluidic chip with an X-shaped micro-channel, a fluid delivery system and a micro-PIV system as shown in Fig. 2.

An X-shaped micro-channel was manufactured on a plastic substrate by precision milling. Besides, a glass cover containing four holes with a thickness of 3 mm was made. To bond the glass cover and the plastic substrate together by adhesive finished preparing the microfluidic channel.

For the fluid delivery system, a syringe pump (KD Scientific Inc.) with two syringes was employed. Polyethylene tubing is used to deliver liquid into the X-shaped channel and to guide the waste streams out of the channel. The experimental conditions are summarized in Table 1.

Besides, the micro-PIV system consists of a microscope lens, a CCD camera, and a quartz halogen illuminator (Dolan-Jenner Industries, 170D) and image post processing system. The images are captured by the Kodak ES1.0 CCD with 1008(H)×1018(V) pixels. Its repetition rate of CCD is 15 Hz. The field-of-view of the experimental field is 2.33 × 2.35 mm^2.

The test rig is filled with seeding particles with diameter of 2 μm for a dense seeding distribution. The generated particle corresponds to 0.87 pixels in the image. After capturing the experimental images by the CCD camera, a pair of images is analyzed to identify the flow field and to calculate the flow velocity based on the particle displacement. The size of the interrogation window is 32 × 32 pixels and the vectors are displayed per 6 or 8 pixels. The illustrations of the internal flow of test section are presented by averaging 100 results for further reduction.

Table 1 Experimental conditions of the present study

Depth of the channel	1.0 mm
Width of the main channel	1.0 mm
Width of the branches	0.5 mm
Inlet flow rates (mL/min) (inlet 1, inlet 2)	(0.1, 0.1) (3.0, 3.0) (0.1, 0.347)
Liquid used	18.3Ω-cm Millipore water
Seeding particle diameter	2 μm

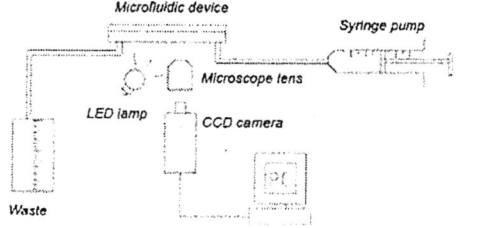

(a) Schematic diagram of the present experimental setup

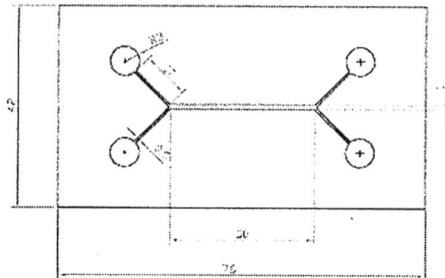

(b) A schematic diagram of the present X-shaped microfluidic channel

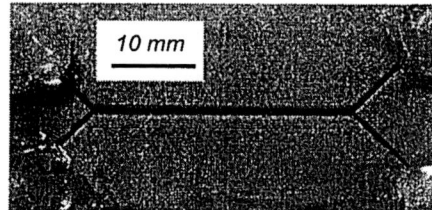

(c) Picture of the present X-shaped microfluidic channel

Fig. 2. Schematic diagram of the present experiment

III. RESULTS AND DISCUSSIONS

A total of three test sections were made and tested, including both forks and the middle region of the X-shaped microchannel.

Figure 3 shows that both streams flow smoothly at the intersection without any mixing when both inlets were fed at the flow rate of 0.1 mL/min. Therefore, one could observe that the interface of both streams lies at the centerline of the main channel as shown in Fig. 4.

In fact, the two parallel streams are laminar in the microchannel for all operations since the maximum Reynolds number was around 67 at the inlet. Consequently, the interface of these two streams is so smooth without any mixing in such microchannel.

Because the performance of such microfluidic fuel cells would usually be limited at the cathode due to the low oxygen concentration [8], the different flow rates fed at anode and cathode would be an option for better performance. Therefore, the above condition was also performed in the present study.

One could observe that the higher flow rate was fed through the lower inlet and resulted in the higher velocity intensity as shown in Fig. 5(a). However, the flow has become fully developed in the middle region of the channel and caused a less velocity intensity near both walls as shown in Fig. 5(b). Finally, the stream leaves the microchannel through two distinct branches almost uniformly.

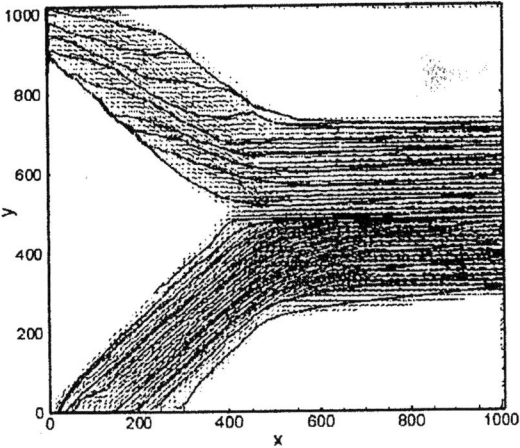

Fig. 3. The flow pattern at the inlets of the X-shaped microchannel (0.1 ml/min, 0.1 ml/min)

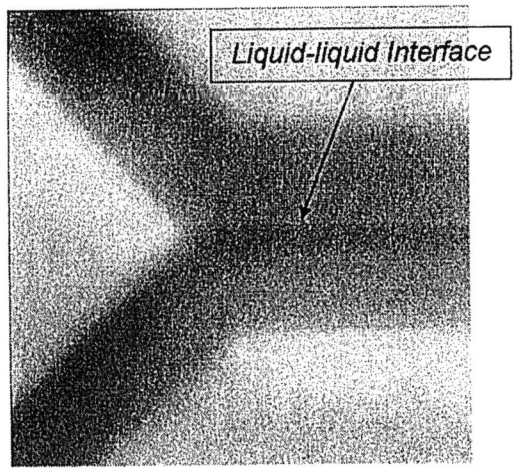

Fig. 4. Visualized result of two streams flow in the X-shaped microchannel

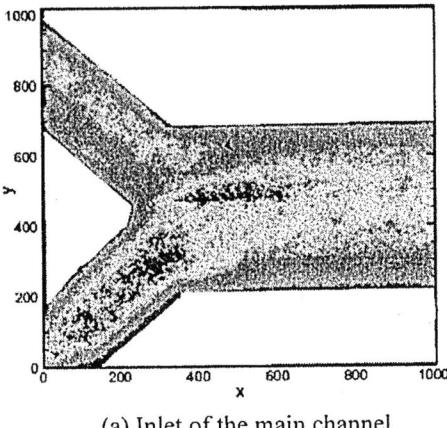

(a) Inlet of the main channel

(b) Middle region of the X-shaped microchannel

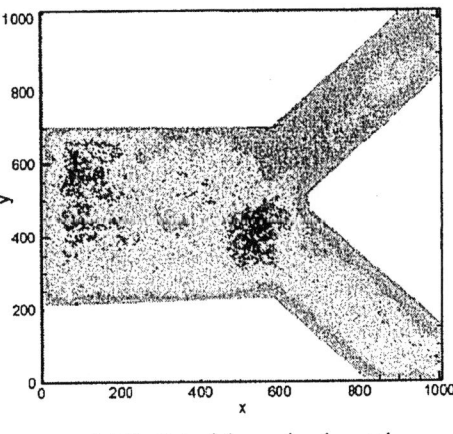

(c) Outlet of the main channel

Fig. 5. The velocity intensity of the analyzed region of the X-shaped microchannel (0.1 ml/min, 0.347 ml/min)

IV. CONCLUSIONS

In order to well design a microfluidic fuel cell, a preliminary study of flow field visualization by micro-PIV in an X-shaped microfluidic channel with two streams was performed. The visualization results show that the two streams for all tests are kept laminar over entire channel even if the different flow rates are fed. Besides, the X-shaped microchannel was confirmed that such design could separate two streams without any mixing at the exit of the main channel for further utilization of these liquids.

Since the flow rate operating range in a microchannel for such application is an important factor, besides, bubbles usually be generated in such process [11]~[12] due to chemical reaction in it, a further study will aim to investigate both the convective and diffusive transport phenomena of a gas/liquid two-phase flow by flow visualization in a microchannel. Moreover, the above X-shaped microchannel will be employed as a microfluidic fuel cell later for power generation tests.

REFERENCES

[1] D. Erickson, and D. Li, "Integrated microfluidic devices," *Analytica Chimica Acta*, vol. 507, pp. 11 26, 2004.

[2] Singhal, V., Garimella, S. V., and Raman, A., "Microscale pumping technologies for microchannel cooling systems," *Appl. Mech. Rev.*, vol. 57, pp. 191-211, January/November 2004.

[3] Müller, M., Müller, C., Gromball, M., Wölfle, W., Menz, Z., "Micro-structured flow field for small fuel cells," *Microsystem Technologies*, vol. 9, pp. 159-162, 2003.

[4] G. Q. Lu, C. Y. Wang, T. J. Yen, X. Zhang, "Development and characterization of a silicon-based micro direct methanol fuel cell," *Electrochimica Acta*, vol. 49, pp. 821-828, 2004.

[5] E. R. Choban, L. J. Markoski, A. Wieckowski, and P. J. A. Kenis, "Microfluidic fuel cell based on laminar flow," *J. Power Sources*, vol. 128, pp. 54 60, 2004.

[6] R. Ferrigno, A. D. Stroock, T. D. Clark, M. Mayer, and G. M. Whitesides, "Membraneless Vanadium Redox Fuel Cell Using Laminar Flow," *J. Am. Chem. Soc.*, vol. 124, pp. 12930-12931, 2002.

[7] A. D. Bazylak, D. Sinton, and N. Djilali, "Improved fuel utilization in microfluidic fuel cells: A computational study," *J. Power Sources*, vol. 143, pp. 57 66, 2005.

[8] E. R .Choban, P. Waszczuk, and P. J. A .Kenis, "Characterization of Limiting Factors in Laminar Flow-Based Membraneless Microfuel Cells," *Electrochem. Solid- State Lett.*, vol. 8, pp. A348-A352, 2005.

[9] J. L. Cohen, D. A. Westly, A. Pechenik, and H. D. Abruna, "Fabrication and preliminary testing of a planar membraneless microchannel fuel cell," *J. Power Sources*, vol. 139, pp. 96 105, 2005.

[10] J. L. Cohen, D. J. Volpe, D. A. Westly, A. Pechenik, and H. D. Abruna, "A Dual Electrolyte H2/O2 Planar Membraneless Microchannel Fuel Cell System with Open Circuit Potentials in Excess of 1.4 V," *Langmuir*, vol. 21, pp. 3544 ~3550, 2005.

[11] K. Sundmacher, and K. Scott, "Direct methanol polymer electrolyte fuel cell: Analysis of charge and mass transfer in the vapor-liquid-solid system," *Chem. Eng. Sci.*, vol. 54, pp. 2927-2936, 1999.

[12] S. Hasegawa, K. Shimotani, K. Kishi, H. Watanabe, "Electricity Generation from Decomposition of Hydrogen Peroxide," *Electrochem. Solid- State Lett.*, vol. 8, pp. A119-A121, 2005.

Mechanical Strain Measurements Using Semiconductor Piezoresistive Material

*Ahmed A. S. Mohammed, *Walied A. Moussa, and *,**Edmond Lou

*University of Alberta, Edmonton, Alberta, Canada
shehata@ualberta.ca, walied.moussa@ualberta.ca and edmond_lou@shaw.ca

Abstract—In this article, the design of a MEMS-based strain sensor has been introduced. This design has high sensitivity, low power consumption compared with the commercially available thin-foil strain gauges, and high absolute resolution. All of these are at high signal stability over a wide temperature range. The piezoresistivity theory, the microfabrication process flow and the finite element simulation have been introduced to provide guidelines for the sensor design process.

I. INTRODUCTION

Silicon strain sensors have many excellent characteristics, such as low power consumption, small size, low hysteresis, high sensitivity and the possibility of monolithic integration [1]. In semiconductor strain gauges, the resistivity changes with strain, along with the dimensions. The net result is a much larger gauge factor than is possible in metal gauges [2]. However, special attention should be paid when designing semiconductor strain gauges to overcome certain challenges. Among these challenges are the temperature dependence of their material properties [3], the need for novel fabrication processes to realize these sensors at high standards of accuracy [4] and the need to account for the signal loss due to signal transfer through the structural layers. Furthermore, the noise effect needs to be considered [5]. The objective from this paper is to introduce the basic guidelines in order to design and fabricate a MEMS-based piezoresistive strain sensor with high sensitivity and high signal stability over a wide temperature range (±50°C).

II. SENSOR DESIGN AND WORKING PRINCIPLE

Figure1 depicts the sensor design showing the sensing unit and microbridge. Surface strain is transferred, from the strained surface, through the bonding layer to the silicon substrate lower surface. Then, the strain is transferred from the silicon substrate bottom surface to the upper surface. Four piezoresistive elements are connected in a full-bridge configuration to have a level of signal magnification. The deformation of the silicon substrate is measured directly from the electrical resistivity change in form of offset voltage.

III. PIEZORESISTIVITY THEORY

When strain (ε) is applied on a conductor, its length (L), cross-sectional area (A) and electrical resistivity (ρ_o) change causing a normalized resistance change of

$$\frac{\Delta R}{R} = (1 + 2\upsilon + \pi Y)\varepsilon \qquad (1)$$

The constant ($1+2\upsilon+\pi Y$) is called piezoresistive gauge factor (G). In metallic materials, the geometric change terms ($1+2\upsilon$) are dominating while in semiconductors, the last term (πY) is of a greater value. According to the piezoresistivity theory [2] and employing the reduced index notation, the normalized resistance change in the off-axis direction cosines (l', m', n') orientation, taking into account that $\pi_{11}=\pi_{22}=\pi_{33}$, $\pi_{44}=\pi_{55}=\pi_{66}$ and $\pi_{12}=\pi_{13}=\pi_{23}$, can be expressed as

$$\frac{\Delta R}{R} = \left(\pi'_{1i}\sigma'_i\right)l'^2 + \left(\pi'_{2i}\sigma'_\alpha\right)m'^2 + \left(\pi'_{3i}\sigma'_i\right)n'^2 + 2\left(\pi'_{4i}\sigma'_i\right)l'n'$$
$$+2\left(\pi'_{5i}\sigma'_i\right)m'n' + 2\left(\pi'_{6i}\sigma'_i\right)l'm' + \left[\alpha_1 T + \alpha_2 T^2 + ...\right] \quad (2)$$

where π_{ij}, $(\alpha_1, \alpha_2,...)$ and T are the on-axis piezoresistive coefficients, the temperature coefficients of resistance and the difference between the reference temperature (T_{ref}) and the working temperature (T_w)

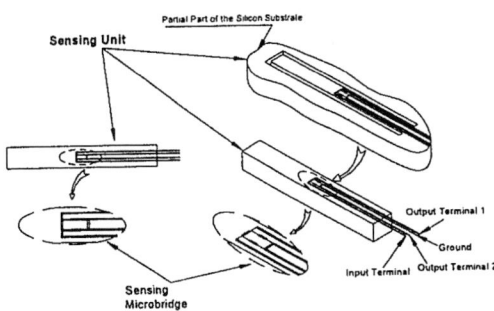

Figure 1: Sensing unit and the microbridge

IV. FINITE ELEMENT MODEL (FEM)

A finite element model, shown in figure2, has been constructed, using ANSYS10.0® software. The strained surface, bonding layer, silicon carrier and piezoresistors were modeled taking into account the isotropy or anisotropy of each layer. The load was applied as a constant displacement on the strained surface edges. Moreover, the boundary conditions' effect has been isolated. Furthermore, the fabrication parameters and the temperature effects have been investigated. The simulation results have been employed to calculate the expected sensor noise and hence its resolution.

V. MICROFABRICATION PROCESS FLOW

The microfabrication process starts by wafers' cleaning in piranha. This step is followed by growing thermal oxide. Next, the first mask is patterned to define the piezore-

1-4244-0899-7/06/$20.00 ©2006 IEEE.

sistors' locations along with the alignment marks. Then, the thermal oxide is etched using buffered oxide etch (BOE) to etch out the locations of the piezoresistors. Boron ion implantation is then performed. A subsequent annealing step follows the ion implantation process. The masking oxide layer is then removed by another BOE step. Aluminum layer is then sputtered to serve in the metallization of the four piezoresistors. Subsequently, the aluminum layer is patterned and etched.

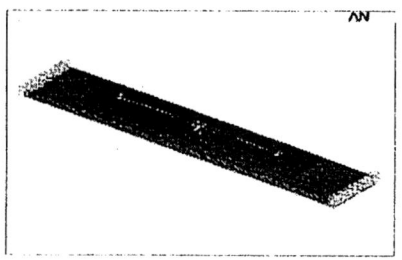

Figure 2: Finite element model

VI. RESULTS AND DISCUSSION

The four piezoresistors are connected to magnify the sensor output signal as well as to help in temperature effect cancellation. Figure3 illustrates the sensor resolution dependence on the working temperature for different doping levels. Figure4 illustrates the sensor resolution dependence on the doping levels at different temperatures.

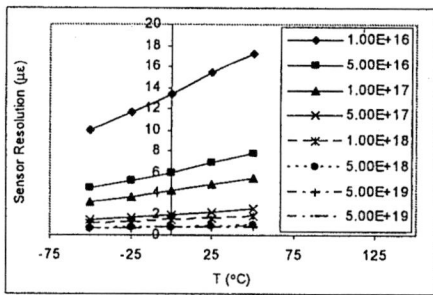

Figure 3: Resolution dependence on temperature

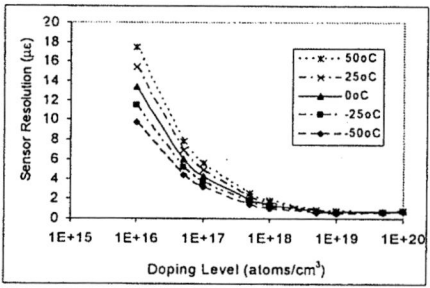

Figure 4: Resolution dependence on doping level

Johnson and 1/f are the most commonly noise sources that affect the piezoresistive sensors [6,7]. Johnson noise is the fundamental performance limit, set by the thermal energy of the carriers in a resistor, and dependent only on the resistance and the working temperature. 1/f noise, on the other hand, affects all resistors at low frequencies due to conductance fluctuations. It is considered originated from the process variables. Therefore, it could be avoided. Moreover, the current which flows in the device presents noise whose power spectral density at low frequency has a divergent behavior. It is noted, from figures3 and 4, that both noise sources are reduced for heavily doped piezoresistors. However, sensitivity considerations favor lightly doped piezoresistors. Moreover, annealing reduces 1/f noise, but causes loss in sensitivity due to dopant diffusion.

VII. CONCLUSION

A MEMS-based piezoresistive strain sensor has been introduced. The working principle and the fabrication process flow are discussed. The main challenges in the piezoresistive strain sensors are the signal temperature dependence and the signal loss. It has been demonstrated that doping concentration and annealing conditions are the most critical parameters that control the performance of piezoresistive sensors. It goes without saying that packaging is one of the most important issues that are facing the commercialization of MEMS piezoresistive sensors.

ACKNOWLEDGEMENTS

This work was supported by Alberta Ingenuity Fund and Alberta CIHR Training Program in Bone and Joint Health, NSERC CRD grant and Syncrude Canada Ltd. The authors would like to thank these organizations for their generous support.

REFERENCES

[1] D. López, R. S. Decca, E. Fischbach, and D. E. Krause,"MEMS-Based Force Sensor: Design and Applications", *Bell Labs Technical Journal*, vol.10, no.3, pp.61–80, 2005.

[2] C.S. Smith, Piezoresistance effect in silicon and germanium, Phys. Rev., vol.94, no.1 pp.42-49, 1954.

[3] Y. Kanda, A graphical representation of the piezoresistive coefficients in silicon, *IEEE Trans.* Electron Devices, vol.29, pp.64-70, January 1982.

[4] C. C. Chang, C. T. Lieu and M. K. Hsich,"Study of the fabrication of a silicon pressure sensor", *Int. J. Electronics*, vol.82, no.3, pp. 295-302, 1997.

[5] B. Bae, B. R. Flachsbart, K. Park and M. A. Shannon,"Design optimization of a piezoresistive pressure sensor considering the output signal-to-noise ratio", *J. Micromech. Microeng.*, vol.14, pp.1597–1607, 2004.

[6] J. A. Harley and T. W. Kenny,"1/F Noise Considerations for the Design and Process Optimization of Piezoresistive Cantilevers", *J. MEMS*, vol. 9, no.2, pp.226-235, JUNE 2000.

[7] H. Nyquist, "Thermal agitation of electric charge in conductors," *Phys. Rev.*, vol. 32, pp. 110-113, 1928.

A New Simple Model for the Single-Electron Transistor (SET)

M. Y. A. Ismail, *Student member, IEEE*, and R. A. AbdelRassoul, *SM, IEEE*

Abstract— We present a new model for simulating the I-V characteristics of a single-electron transistor (SET) at the steady-state mode based on a reduced master equation (ME) method. The model is accurate, fast and less numerically intensive. A comparison is made between SET simulation using our model and that generated by the model based on full master equation method of the quantum transport (QT) research group at Delft University, which considers all possible charge states in the tunnel junction. The comparison shows that results of our fast model are in excellent agreement with QT's results at low bias conditions, but show some deviation at large bias. the footnote at the bottom of this column.

I. INTRODUCTION

The continuing scaling down and miniaturization of CMOS devices has led researchers now to build new microelectronic devices with very small dimensions (nanotechnology), whose behavior will be interpreted based on quantum mechanics. The single-electron transistor (SET) is one of these devices.

The idea of using the SET device was first suggested in 1985 by Dmitri Averin and Konstantin Likharev [1],[2] who were studying the properties of a well known electronic device which is the tunnel junction under the conditions of minimizing its dimensions to the nanometer range. Two years later, Theodore Fulton and Gerald Dolan at Bell Labs fabricated such device and demonstrated its operation [3,4]. These junctions can be implemented using a variety of materials, such as metal-insulator-metal structures, GaAs quantum dots, silicon structures, large molecules with conducting cores, etc. The early types of SET devices were mainly metallic.

The tunnel junction which is the main component of the SET device consists of two layers of conducting material separated by a layer of an insulating material which is as thin as about 1 nm. If the insulating material was thick, the tunnel junction would become a conventional capacitor, i.e. if a voltage is applied on the junction a +Q charge would be generated on one side of the junction and a −Q charge on the other side. But in the tunnel junction and according to the laws of quantum mechanics, each electron has a chance to pass (tunnel) through the tunnel junction and according to its direction of jump, an increase or decrease for the initial charge Q by an amount of e will occur. In the same time, a change for the voltage across the junction by a value of e/C will also occur.

Previous work on modeling of single-electron devices is based on (i) the Monte Carlo modeling, with MOSES [5] and SIMON [6] typical simulators, (ii) the Master Equation modeling, with SENECA [7] as the simulator, and (iii) the Deterministic modeling. A compact analytical model for asymmetric single-electron tunneling transistors can be found in [8].

II. THEORY OF SINGLE-ELECTRON TRANSISTOR (SET) OPERATION

One of the popular methods used in modeling of SET devices is originating from applying the Markov process which has the following postulate: electrons do not posses any memory when tunneling through the junction [9]. Accordingly, the tunnel rates will depend only on the system state. Mathematically, such a system can be described by the following equation :

$$\frac{\partial P_i(t)}{\partial t} = \sum_{j \neq i} [\Gamma_{ij} P_j(t) - \Gamma_{ji} P_i(t)] \qquad (1)$$

Where Γ_{ij} is the tunnel rate from state j to state i and $P_i(t)$ is the time dependent occupation probability of state i. The basic idea standing behind equation (1) is to write down all possible tunnel rates for the different states in the many electron system. After writing down the rates along with their corresponding probabilities (considering that tunnel rates can be easily calculated from the orthodox theory), we solve the system of equations for finding the probabilities of the different states.

In order to have a full master equation solution, we have to consider all possible charge states in the tunnel junction. For sure this technique will result in heavy numerical calculations which will consume long time and needs large capacity of memory. However one of the solutions is to adopt a steady state three-state model [10] for the capacitively-coupled SET shown in Fig. (1). However the only constraint for this method that it only works well only if the SET can be regarded as a separate component in the circuit.

This condition can be easily met if SET devices are connected to circuit nodes that have large capacitances to insure that the charging energy ($e^2/2C$) per electron is very small on the external interfaces and accordingly we can easily adopt their method.

M. Y. A. Ismail was with the Arab Academy for Science and Technology, Alexandria 21937 EGYPT. He is now with Qatar Steel, Qatar.

R. A. AbdelRassoul is with the Arab Academy for Science and Technology, Alexandria 21937 EGYPT. (e-mail: Roshdy@gmail.com)

1-4244-0899-7/06/$20.00 ©2006 IEEE.

III. THE NEW MASTER EQUATION (ME) – BASED MODEL

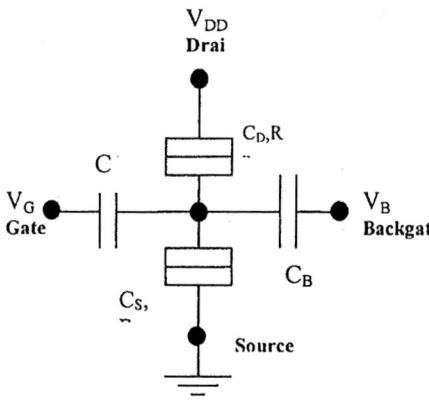

Fig. (1) The equivalent circuit of a capacitively-coupled single-electron transistor (SET)

Now we will go through the model details. We start with writing down the steady state master equations for three charge states (n-1,n,n+1) only, equation (2). Fig (2) represents the state transition diagram for limited three charge states(n-1,n,n+1):

$$\begin{bmatrix} -\Gamma(n/n-1) & \Gamma(n-1/n) & 0) \\ \Gamma(n/n-1) & -[\Gamma(n+1/n)+\Gamma(n-1/n)] & \Gamma(n/n+1) \\ 0 & \Gamma(n+1/n) & -\Gamma(n/n+1) \end{bmatrix}\begin{bmatrix} p_{n-1} \\ p_n \\ p_{n+1} \end{bmatrix}=0 \quad (2)$$

where $\Gamma(n-1/n)$: is the tunneling rate from state n to state n-1, and p_n: is the occupation probability of the state n.

From equation (2) we get all tunnel rates (from the orthodox theory), but we do not know the occupation probability for these rates. Writing the resulting equations in matrix form, we get:

$$V_0 = \frac{1}{C}[V_G C_G + V_B C_B + V_{DD} C_D + ne] \quad (3)$$

Eq.(3) represents the solution for occupation probability for the three states.

$$\Gamma(\Delta F) = \frac{\Delta F}{e^2 R(1-e^{-\Delta F/kT})} \quad (4)$$

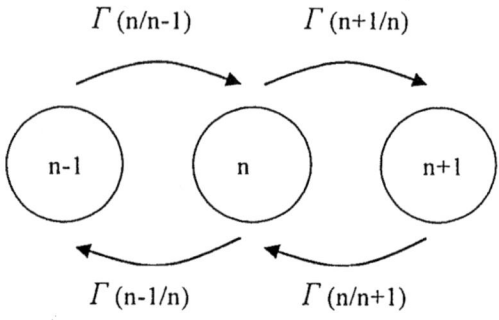

Fig. (2). State transition diagram for SET device considering only three charge states (n-1,n,n+1)

Now we need to calculate the tunnel rates. Based on the orthodox theory the tunnel rate is calculated by eq.(4) [11], where ΔF is the change in the free energy in the tunnel junction before and after the tunneling event, R is the resistance of the tunnel junction, k is Boltzmann constant and T is the absolute temperature. Now to calculate ΔF which is described by eq.(5) [7]:

$$\Delta F = \frac{e^2}{2C} \pm e(\Delta V) \quad (5)$$

where C is the summation of all capacitances, ΔV is the voltage difference across the tunnel junction under observation. From equations (4) and (5), it is clear that we need to calculate (ΔV) first to get all other values. Fortunately ΔV can be easily calculated from basic classical electronics. Applying Kirchoff's Current Law at the island node to get:

$$(V_G - V_0)C_G + (V_B - V_0)C_B + (V_{DD} - V_0)C_D - V_0 C_S + ne = 0 \quad (6)$$

Rearranging, we get:

$$V_0 = \frac{1}{C}[V_G C_G + V_B C_B + V_{DD} C_D + ne] \quad (7)$$

where

$$C = C_D + C_S + C_G + C_B \quad (8)$$

The voltage difference across the tunnel junction is:

$$\Delta V_D = V_{DD} - V_0 = \frac{1}{C}[V_{DD}(C-C_D) - V_G C_G - V_B C_B - ne] \quad (9)$$

Substituting from eq.(9) in eq.(5), we get:

$$\Delta F_D = \frac{e}{C}[e/2 + V_{DD}(C-C_D) - V_G C_G - V_B C_B - ne] \quad (10)$$

Thus substituting from eq.(16) in eq.(10) for the value of change of free energy, we get the tunnel rate for electron through the Drain junction. Similarly, we can get:

$$\Delta F_S = \frac{e}{C}[e/2 + V_{DD}C_D + V_G C_G + V_B C_B + ne] \qquad (11)$$

The net steady state current flowing in drain junction is equal to that of the source one, thus the steady state current is:

$$I = e(P_{n-1}\Gamma(n/n-1) + P_n\Gamma(n+1/n)) \qquad (12)$$

Eq. (12) is calculated based on tunneling through the drain junction. For tunneling through the source junction:

$$I = e(P_n\Gamma(n-1/n) + P_{n+1}\Gamma(n/n+1)) \qquad (13)$$

The currents calculated from eq.(12) and eq.(13) are both equal.

IV. Simulation Results and Comparison with the Quantum-Transport (QT) Model

In order to check the validity of our proposed fast model we selected as a benchmark the modeling software available on the web developed by the "Quantum Transport" (QT) research group in Delft University, which is based on full-master equation method, and which considers 200 different charge states during the program execution [12,13].
The comparison was made with 7 different parameters of a single-electron transistor (SET), as shown in Fig. 3 – 5.
The results of our proposed fast model and those of the QT model are exactly matching . We made the comparison with different tunneling resistances, R_d, different back-gate capacitance, C_b, different gate voltage bias, V_g, (Fig. 3 and Fig. 5), and finally at different temperatures, T, (Fig. 4 and Fig. 5). However we found that at high bias voltage (V_{dd}) there is a larger deviation between our fast model and the QT's. Fig. 10 shows the combined I-V characteristics for all cases for the fast model.

V. Conclusion

Our proposed fast model gives exactly the same results at low bias voltage conditions when compared to QT's model. This means that our fast model which needs less computational mathematics and accordingly less memory and less CPU usage will lead to the same results generated by QT model which depends on full master-equation method. However we still need to improve our fast model in order to achieve better agreement at high bias conditions. Very good agreement is obtained between the new fast model and the quantum-transport (QT) model.

VI. References

[1] K. K. Likharev and T. Claeson,, "Single Electronics", Scientific American, vol. 266, no.. 6, pp. 80-85, June 1992.

[2] A. N. Korotkov, "Coulomb Blockade and Digital Single-Electron Devices" http://arxiv.org/abs/cond-mat/9602165

[3] M. H. Devoret and C. Glattli, "Single-electron transistors", "http: // physicsweb.org / articles / world/11/9/7/1".

[4] M. Roukes, "Plenty of Room Indeed", Scientific American-September 2001, can be downloaded from "www.its.caltech.edu / ~nano/papers / SciAm-Sep01.pdf".

[5] R. H. Chen et al., MS-DOS version of MOSES 1, http: // hana.physics. sunysb.edu /set/software/index.html.

[6] C. Wasshuber, H. Kosina and S. Selberherr, "SIMON - a simulator for single-electron tunnel devices and circuits", IEEE Transactions on Computer-Aided Design, vol. 16, no.9, pp.937-944, September 1997.

[7] L. R. C. Fonseca, A. N. Korotkov, K. K. Likharev and A. A. Odintsov, " A Numerical Study of the dynamics and statistics of single-electron systems", J. Appl. Phys., vol. 78, no. 5, pp.3238-3251, September 1995.

[8] H. Inokawa,and Y. Takahashi, "A Compact Analytical Model for Asymmetric Single-Electron Tunneling Transistors ", IEEE Transactions on Electron Devices , vol. 50, no. 2, February 2003.

[9] C. Wasshuber, Computational Single-Electronics (Computational Microelectronics series), Springer-Verlag, Wien, 2001.

[10] C. H. Hu, J. F. Jiang and Q. Y. Cai, "An Improved Three State Master Equation Model for Capacitively Coupled Single-Electron Transistor", IEEE-NANO 2002 Conference.

[11] K. K. Likharev, "Single Electron Devices and Their Applications", Proceedings of The IEEE, Vol. 87, No.4, pp.606-632, April 1999.

[12] http://qt.tn.tudelft.nl/~hadley/set/asymIV/ SETIV.html .

[13] http://qt.tn.tudelft.nl/research/set/setnets/ setnets.html.

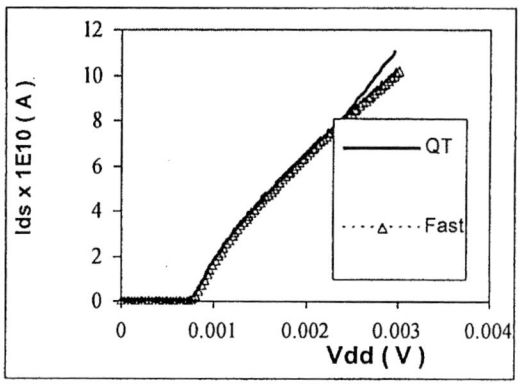

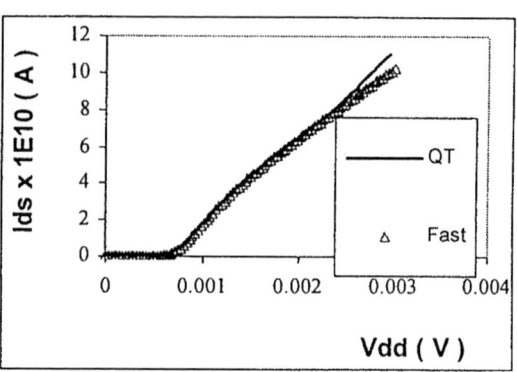

Fig. (3) I-V curve for SET device with the following parameters: Cd=1e-16;Cs=Cd;Cg=1e-18; Cb=Cg; Vg=0; Vb=0; Rd=1e6; Rs=Rd; T=0.1.

Fig. (4) I-V curve for SET device with the following parameters: Cd=1e-16; Cs=Cd; Cg=1e-18; Cb=0; Vg=0.009; Vb=0; Rd=1e6; Rs=Rd; T=0.3 .

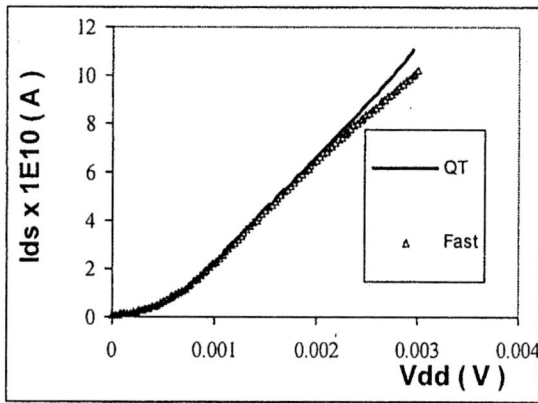

Fig. (5) I-V curve for SET device with the following parameters: Cd=1e-16; Cs=Cd; Cg=1e-18; Cb=0; Vg=0.009; Vb=0; Rd=1e6; Rs=Rd; T=0.1

An Advanced Model for Dopant Diffusion in Polycrystalline silicon during rapid thermal annealing

S. Abadli and F. Mansour

Abstract—We have investigated and modelled the diffusion of boron implanted into polycrystalline silicon. A one-dimensional two stream diffusion model adapted to the granular structure of polysilicon and to the effects of the strong-concentrations has been developed. This model includes dopant clustering in grains as well as in grain boundaries. The grains-growth and energy barrier height are coupled with the dopant diffusion coefficients and the process temperature based on thermodynamic concepts. The simulation well reproduces the experimental profiles when crystallisation and clustering are considered. The trapping-emission mechanism between grains and grain boundaries and segregation are the major effects during annealing process.

I. INTRODUCTION

THE compatibility of polycrystalline silicon thin films with subsequent thermal processing allows its efficient integration into advanced integrated-circuits and permits fabrication of new devices structures [1], [2]. The heavily doped polysilicon is currently the more used gates material. Ion implantation remains the most used technique of doping; it allows a good control of diffusion profiles. However, the implanted dopant is generally electrically inactive [3], and the energetic ions create a large concentration of defects or damages that degrade the device characteristics [3], [4]. Thermal post-implantation annealing is essential to treat the samples of its defects and to allow the implanted ions to take positions where they will be electrically active and able to exchange charges with the silicon atoms. During this process, the presence of high self-interstitial clusters causes transient enhanced diffusion (TED) of the dopant atoms [4], [5]. This poses a major problem for the manufacture of micro-electronics components and advanced integrated circuits.

At strong concentrations of impurities, solubility solid limit can be exceeded and the doping excess precipitates and forms inactive and immobile clusters [6], [7]. Moreover, dopant forms electrically inactive and immobile clusters even at concentrations far below the solid solubility limit under the supersaturation of self-interstiels [8]. During thermal annealing, dopant complex redistribution in polysilicon is strongly affected by the morphological structure of polysilicon. An additional complexity is that the morphological structure of polysilicon changes during annealing, with grains growing in size. This means that the simulation of the diffusion profiles is more and more complex and very difficult. In this work, our aim is to develop a fundamental understanding of the boron transfer mechanisms inside polysilicon and developing a model for the process. We wish to more finely understand the role of the grain boundary and that of the grains growing in size in the TED during the the rapid thermal post-implantation annealing (RTA). A theoretical two-stream diffusion model adapted to the granular structure of polysilicon and to the effects of the strong concentrations has been developed. This model consists of two coupled partial differential equations of diffusion. The first is associated to the diffusion in the grains; the second is associated to the diffusion in the grain boundaries. These two equations are coupled by a term which represents the transfer and the opposite transfer between the grains and the grain boundaries by associating effects related to the strong-concentrations and that of the segregation phenomenon. The adjustment of the simulated profiles with the experimental SIMS profiles for different RTA temperatures (700, 750 and 800°C) for 1 min allowed the validation of this model.

II. EXPERIMENTAL DETAILS

The studied samples consist of 335-nm-thickness layers of amorphious silicon obtained at 465°C by low pressure chemical vapor deposition (LPCVD) of disilane Si_2H_6, under pressure of 200 mTorr. After a standard clean, the films are deposited on oxidized monocrystalline silicon substrates (P-type, (111), 120 nm of thermal oxide SiO_2). They are then boron implanted with a dose of 4×10^{15} atoms/cm^2 at an energy of 15 keV [9]. In order to avoid long redistributions, post-implantation annealing was carried out at a relatively low temperatures (700, 750 and 800°C) and short times of 1 min. The experimental doping profiles have been obtained by secondary ion mass spectrometry (SIMS) with a CAMECA IMS4F measuring device [9].

By the use of the analytical Gaussian expression known by the three following parameters: the ion implantation dose Q_d, the projected range R_p, and the straggle ΔR_p, given by [10]:

$$C(x) = \frac{Q_d}{\sqrt{2\pi}\Delta R_p} \exp\left(-\frac{(x-R_p)^2}{2\Delta R_p^2}\right) \quad (1)$$

we can easily simulate the total initial redistribution profile just after ion implantation. Figure 1, shows the good superposition of simulated profile and SIMS profile before

S. Abadli and F. Mansour are with the department of Electronics, University Mentouri Route d'Ain EL-Bey Constantine, Constantine 25000, Algeria (e-mail: abadli_com_art@yahoo.fr).

1-4244-0899-7/06/$20.00 ©2006 IEEE.

annealing. This will be used as reference or initial condition during theoretical simulation step.

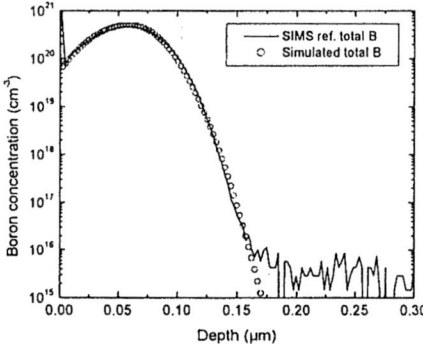

Fig. 1. Superposition of the simulated profile with the experimental SIMS profile before annealing

From the study of Solmi *et al.* [11], we can tell that the implantation dose of 4×10^{15} atoms/cm^2 introduce the exceeding of the solubility solid limit at these annealing temperatures. In this case, the doping excess precipitates and forms inactive and immobile clusters [4], [7]. In the reality, the state of the dopant related to this parameter was not given in a clear and single way in the literature. The ones suggests that dopant segregate to the grain boundaries [6], [12], while the others propose the possibility of clustering even at concentrations far below the solubility limit under the supersaturation of self-interstitials [7], [13].

III. MODEL

A If lot of models had been proposed for the study and the simulation of the complex mechanisms of boron redistribution and activation in single-crystal and polycrystalline silicon during the thermal post-implantation annealing [4], [9], and [14]. By using the ideas of each model, we describe now a more detailed model, adapted to the granular structure of polysilicon and to the effects of very strong concentrations. The diffusion is controlled by two differential partial equations coupled by a term representing the dopant transfer between the grains and the grain boundaries. The total concentration will be divided between the grains and the grains boundaries. The effects connected to the heavily concentrations were combined with the dopant diffusion coefficients in the grains and the grain boundaries. For a one-dimensional diffusion this model takes the following form:

$$\frac{\partial C_g}{\partial t} = \frac{\partial}{\partial x}\left(D_g^{eff}\frac{\partial C_g}{\partial x}\right) - k_t^{eff}\left(C_g - \frac{C_{gb}}{k_{seg}}\right) \quad (2)$$

$$\frac{\partial C_{gb}}{\partial t} = \frac{\partial}{\partial x}\left(D_{gb}^{eff}\frac{\partial C_{gb}}{\partial x} + \frac{D_{gb}^{eff}}{L_g}C_{gb}\frac{\partial L_g}{\partial x}\right) + k_t^{eff}\left(C_g - \frac{C_{gb}}{k_{seg}}\right) \quad (3)$$

C_{gb} is the dopant concentration in the grain boundaries and C_g is the dopant concentration inside the grains. The total concentration in the studied layers is thus, the sum of the two concentrations ($C_{total}=C_g+C_{gb}$). D_g^{eff} and D_{gb}^{eff} are respectively the effective diffusion coefficients in the grains and the grain boundaries. They are identified by:

$$D_g^{eff}=D_i\frac{1+\beta(p/n_i)}{1+\beta}\left(1+\frac{C_g}{\sqrt{C_g^2+4n_i^2}}\right)\left(1+m\left(\frac{C_g}{C_{sol}}\right)^{2m}\right)^{-1} \quad (4)$$

$$D_{gb}^{eff}=D_iF_a\left(1+\frac{C_{gb}}{\sqrt{C_{gb}^2+4n_i^2}}\right)\left(1-\exp\left(\frac{-E_b}{KT}\right)\right) \quad (5)$$

$$D_i=D_0\exp\left(\frac{-E_a}{KT}\right) \quad (6)$$

D_i is the intrinsic diffusivity in single-crystal silicon, its value varies exponentially with the temperature T and the activation energy E_a associated with the diffusion process. In this study, we took the value of E_a=3.46eV [15]. D_0 is the diffusivity pre-exponential factor. The effective diffusion coefficient in the grains is greatly linked to the high concentration effects. It depends on the vacancies concentrations in the grains, under the effect of the ratio of the diffusivity induced by the charged vacancies on the global diffusivity induced by the neutral vacancies $\beta=D_i^+/D_i^0$ [9], [15], of holes concentration p, and of the intrinsic concentration n_i. D_g^{eff} depends also, of the doping solubility solid limit C_{sol}, as well as, of the clusters size by the maximum number m of self-interstitials sites occupied by clusters [4], [14]. Concerning the effective diffusion coefficient in the grain boundaries, it is well controlled by the trapping and the segregation to the grain boundaries [16]. These two effects are obviously related to the energy barrier E_b to the grain boundaries, it even, depends on the impurities concentration, the average size of the grains, and the traps density [17], [18]. F_a is a pre-exponential factor for the adjustment of the intrinsic diffysivity in polycrystalline silicon thin films.

The coupling between the diffusion equations (2) and (3) is ensured by the term representing the effective dopants transfer from the grains to the grain boundaries and vice versa. The net effective transfer rate is given by [16], [19]:

$$k_t^{eff}=\frac{D_g}{L_g}\left(\frac{4}{L_g}+\frac{1}{2\sqrt{D_gt}}\right)+\frac{2\alpha}{L_g}\frac{\partial L_g}{\partial t} \quad (7)$$

The effective doping transfer rate between the grains and the grain boundaries depends primarily of the change of the grain average diameter L_g during thermal annealing. α is a factor of adjustment; in our work we take α=1 [19]. The grains-growth is proportional to the square root of time [16], [20]:

$$L_g(t)=\sqrt{L_{g0}^2+2\sigma t} \quad (8)$$

$L_g(t)$ represents the average grains size after time t and L_{g0} represents the initial average grains size after deposition of polysilicon layer. σ is a parameter which depends on the grain boundary mobility and the grain boundary energy. Its value depends on the local Fermi level. Concerning the dopant segregation, it is expressed by using the results of Mandurah et al. [21] and swaminathan et al. [22]. It is given by:

$$k_{seg} = 2k_{seg0} \frac{e_{gb}}{L_g} \qquad (9)$$

k_{seg0} is the thermal equilibrium grain boundary segregation coefficient and e_{gb} is the average width of a grain boundary. Segregation coefficient is described by the expression similar to one of Mandurah et al. [21]: $K_{seg0} = K_0 \exp(0.456/KT)$.

IV. RESULTS AND DISCUSSION

The simulated boron diffusion profiles were calculated by means of a realized program; which makes the numerical resolution of the differential equations (2) and (3); while using an implicit method of finite differences. From the good superposition of the simulated profiles and SIMS profiles, we can illustrate the significant role of the grains growing in size and that of the dopant trapping and segregation to the grain boundaries. The adjustment is obtained while varying the following parameters: L_{g0}, $F_a = D_{poly}/D_{mono}$, m, β, k_{seg}. Figures 2 4, show clearly the good adjustment of the simulated profiles with SIMS profiles. In these figures, we notice good matching, particularly for the profiles shoulder that occurs for the boron solubility limit. The good adjustment indicates the validation of this model and confirms that the grins-growth in size play a significant role for the precise determination of the diffusion profiles. We can tell that the boron diffusivity in the grains is not very different to that in the grain boundaries. This is justified by the significant reduction of the diffusion coefficient in the grain boundaries caused by the important trapping-emission mechanism. In effect, the effective diffusion coefficient in grain boundary depends much on E_b, L_g and thus on the density of traps. The vacancies concentrations effect on boron diffusivity depends principally of β, its optimal value obtained after profiles adjustment is 0.13, which is in good agreement with that in the results of Mahamdi et al. [9] and Giroult et al. [23]. The adjustment factor of diffusivity in the grain boundaries given in the theory by the ratio D_{poly}/D_{mono} is approximately 100 in our simulation. This means that the boron diffusivity in the grain boundaries is about 100 times greater than that in the grains. This result is consistent with that obtained by Probst et al [15]. In this work, the number of self-interstitials to be trapped in the boron clusters m, takes the value of $m=2$, led to the best fitting for all the profiles. This m value is approximately similar to that obtained by Uematsu for the monocrystalline silicon [4], [7]. In consequence, the presence of high self-interstitial concentration causes transient enhanced diffusion (TED) of dopant atoms and their precipitation into clusters at a concentration well below the solubility solid limit. This was observed in rigorous manner in the simulated profiles by the shoulder (bump) of the diffusion profiles that occurs with the boron solubility solid limit. This effect was also discussed in research of Pelaz et al. [5], [13] and research of Mahamdi et al. [9].

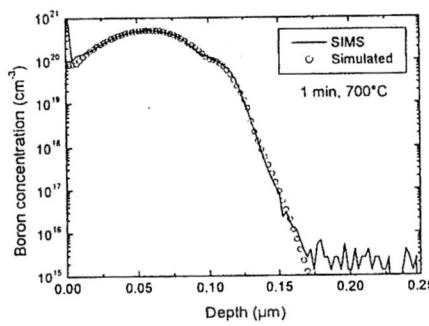

Fig. 2. Superposition of the simulated profile with the experimental SIMS profile after annealing (700°C/1min)

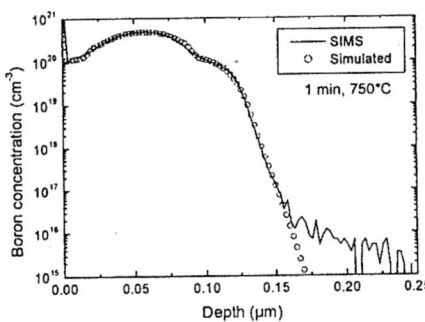

Fig. 3. Superposition of the simulated profile with the experimental SIMS profile after annealing (750°C/1min)

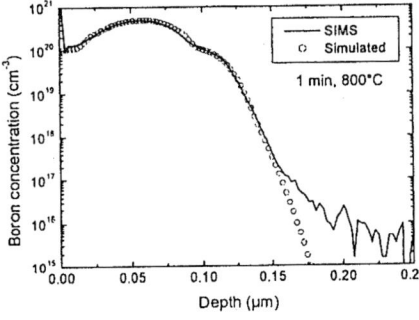

Fig. 4. Superposition of the simulated profile with the experimental SIMS profile after annealing (800°C/1min)

V. CONCLUSIONS

According to the results of this study, grains-growth and clustering effects are two very significant parameters for the precise simulation of the diffusion profiles. In addition, we can note that the boron transient enhanced diffusion (TED) does not depend only on the trapping to the grains boundaries, but also on the clustering effect to the grains, under the effects of the strong concentrations. For all high doses implantation cases the trapping-emission mechanism between grains and grain boundaries and growth of grains are the major effects during thermal annealing process.

REFERENCES

[1] B. Yu, D. H. Ju, W. C. Lee, N. Kepler, T. J. King and C. Hu, "Gate engineering for deep-submicron CMOS transistors," *IEEE, Trans. Electron Devices*, vol. 45, pp. 1253 1262, 1998.

[2] A. J. Walker, S. B. Herner, T. Kumar and En-H. Chen, "On the conduction mechanism in polycrystalline silicon thin-film transistors," *IEEE, Trans. Electron Devices*, vol. 51, No. 11, pp. 1856 1866, 2004.

[3] L. Pelaz, V. C. Venezia, H. J. Gossmann, G. H. Gilmer, A. T. Fiory, and C. S. Rafferty, "Activation and deactivation of implanted B in Si," *Appl. Phys., Lett.*, vol. 75, No. 5, pp. 662 664, 1999.

[4] Masashi Uematsu, " Simulation of high-concentration boron diffusion in silicon during post-implantation annealing," *Jpn. J. Appl. Phys.*, vol. 38, pp. 3433 3439, 1999.

[5] M. Jaraiz, G. H. Gilmer and J. M. Poate, "Atomistic calculations of ion implantation in Si: point defect and transient enhanced diffusion phenomena," *Appl. Phys., Lett.*, vol. 68 No. 3, pp. 409 411, 1996.

[6] H. Schaber, R. V. Criegern and I. Weitzel, " Analysis of polycrystalline silicon diffusion sources by secondary ion mass spectrometry," *J. Appl. Phys.*, vol. 58, No. 11, pp. 4036 4042, 1985.

[7] Masashi Uematsu, " Simulation of clustering and transient enhanced diffusion of boron in silicon," *J. Appl. Phys.*, vol. 84, No. 9, pp. 4781 4787, 1998.

[8] J. R. Pfiester, F. K. Baker, T. C. Mele, H. H. Tseng, P. J. Tobin, J. D. Hayden, J. W. Miller, C. D. Gunderson and L. C. Parillo, "The effects of boron penetration on P^+ polysilicon gated PMOS devices," *IEEE Trans. Electron Devices*, vol. 37, No. 8, pp. 1842 1847, 1990.

[9] R. Mahamdi, F. Mansour, E. Scheid, P.T. Boyer and L. Jalabert, "Boron diffusion and activation during heat treatement in heavily doped polysilicon thin films for P^+ Metal-Oxyde-Semiconductor transistors gates," *Jpn. J. Appl. Phys.*, vol. 40, pp. 6723 6727, 2001.

[10] R. W. Cahn, P. Haasen, and E. J. Kramer, *Electronic structure and properties of semiconductors*, vol. 4, April 1991, pp. 264 275.

[11] S. Solmi, F. Baruffaldi and R. Canteri, "Diffusion of boron in silicon during post-implantation annealing," *J. Appl. Phys.*, vol. 69, No. 4, pp. 2135 2142, 1991.

[12] S. Batra, M. Manning, C. Dennison, A. Sultan, S. Bhattacharya, K. Park, S. Banerjee, M. Lobo, G. Lux, C. Kirschbaum, J. Noberg, T. Smith and B. Mulvaney, "Discontinuity of B-diffusion profiles at the interface of polycrystalline Si and single crystal Si," *J. Appl. Phys.*, vol. 73, No. 8, pp. 3800 3804, 1993.

[13] L. Pelaz, G. H. Gilmer, H. J. Gossmann, C. S. Rafferty, M. Jaraiz and J. Barbella, "B cluster formation and dissolution in Si: A scenario based on atomic modeling," *Appl. Phys., Lett.*, vol. 74, pp. 3657 3660, 1999.

[14] H. Puchner and S. Selberherr, " An advanced model for dopant diffusion in polysilicon," *IEEE, Trans. Electron Devices*, vol. 42, No 10, pp. 1750 1754, 1995.

[15] V. Probst, H. J. Bohm, H. Schaber, H. Oppoler and I. Weitzel, "Analysis of polysilicon diffusion sources," *J. Electrochem Soc.*, vol. 135, No. 3, pp. 671 676, 1988.

[16] A. D. Sadovnikov, "One-dimensional modeling of high concentration boron diffusion in polysilicon-silicon structures," *Solid-State Electronics*, vol. 34, No. 9, pp. 969 975, 1991.

[17] J. Y. W. Seto, "The electrical properties of polycrystalline silicon films," *J. Appl. Phys.*, vol. 46, No. 12, pp. 5247 5254, 1975.

[18] G. Baccarani, B. Ricco and G. Spadini, "Transport properties polycrystalline silicon films," *J. Appl. Phys.*, vol. 49, pp. 5565 5570, 1978.

[19] S. K. Jones and C. Hill, "Modeling dopant diffusion in polysilicon," *Simulation of Semiconductor Devices and processes*, vol. 3, pp. 441 449, 1988.

[20] H.-J. Kim and C.V. Thompson, "Kenetic modeling of grains growth in polycrystalline silicon films doped with phosphorus or boron," *J. Electrochem. Soc.*, vol. 135, pp. 2312 2316, 1988.

[21] M. M. Mandurah, K. C. Saraswat, C. R. Helms and T. I. Kamins, "Dopant segregation in polycrystalline silicon," *J. Appl. Phys.*, vol. 51, No. 11, pp. 5755 5763, 1980.

[22] B. Swaminathan, E. Demoulin, T. W. Sigmon, R. W. Dutton and R. Rif, "Segregation of arsenic to the grain boundaries in polycrystalline silicon," *J. Electrochem Soc.*, vol. 127, No. 10, pp. 2227 2229, 1988.

[23] G. Giroult, A. Nouailhat and M. Gauneau, "Study of Wsi$_2$ / polycrystalline silicon / monocrystalline silicon structure," *Appl. Phys.*, vol. 67, pp. 515 523, 1990.

Growth of Grains Effect on Boron Diffusion in Heavily Implanted Polycrystalline silicon Thin Films

S. Abadli and F. Mansour

Abstract—A one-dimensional two stream diffusion model adapted to the granular structure of polysilicon and to the effects of the strong concentrations has been developed. This model includes dopant clustering in grains as well as in grain boundaries. Growth of grains and energy barrier height are coupled with the dopant diffusion coefficients and the process temperature based on thermodynamic concepts. The adjustment of the simulated profiles with the experimental SIMS profiles for short treatment times ranging between 1 and 30 minutes at temperature of 700°C; allowed the validation of this model. Growth of grains and strong-concentrations phenomena are the major effects during annealing processes. They play a significant role for the precise determination of the diffusion profiles.

I. INTRODUCTION

THE major use of polysilicon thin films in VLSI circuits is as the gate electrodes and one level of interconnections in CMOS integrated circuits. The ability to self-align the gate electrodes of MOS field-effect transistors reduces capacitance and improves circuits speed. In addition to improving circuits speed, the compatibility of polycrystalline silicon thin films with subsequent thermal processing allows its efficient integration into advanced integrated-circuits and permits fabrication of new devices structures [1], [2]. In MOS technology, the heavily doped polysilicon is currently the more used gates material. Ion implantation remains the most used technique of doping; it allows a good control of diffusion profiles. However, the implanted dopant is generally electrically inactive [3], and the energetic ions create a large concentration of defects or damages that degrade the device characteristics [3], [4]. Thermal post-implantation annealing is essential to treat the samples of its defects and to allow the implanted ions to take positions where they will be electrically active and able to exchange charges with the silicon atoms. Therefore, post-implant thermal processing is required to anneal out the damage and to electrically active the dopant. During this process, the presence of high self-interstitial clusters causes transient enhanced diffusion (TED) of the dopant atoms [4], [5]. This poses a major problem for the manufacture of micro-electronics components and advanced integrated circuits; because TED is a limiting factor in the scaling down of device size. Indeed, simulation of TED and activation of boron in polysilicon thin films is essential. At high-concentration of impurities, solubility solid limit can be exceeded and the doping excess precipitates and forms inactive and immobile clusters [6], [7]. Moreover, dopant forms electrically inactive and immobile clusters even at concentrations far below the solid solubility limit under the supersaturation of self-interstitiels [8]. During thermal annealing, dopant complex redistribution in polysilicon is strongly affected by the morphological structure of polysilicon. An additional complexity is that the morphological structure of polysilicon changes during annealing, with grains growing in size. This means that the simulation of the diffusion profiles is more and more complex and difficult.

In this paper, our aim is to develop a fundamental understanding of the boron transfer mechanisms inside polysilicon and developing a model for the process. With the aim of more finely understanding the role of the grain boundary and that of the grains growing in size in the TED during the thermal post-implantation annealing, we propose a theoretical two-stream diffusion model adapted to the granular structure of polysilicon and to the effects of the strong concentrations. This model consists of two coupled partial differential equations of diffusion. The first is associated to the diffusion in the grains; the second is associated to the diffusion in the grain boundaries. These two equations are coupled by a term which represents the transfer and the opposite transfer between the grains and the grain boundaries by associating effects related to the strong concentrations and that of trapping-segregation.

II. EXPERIMENTAL DETAILS

A. Experiment

The studied samples consist of 335-nm-thickness layers of amorphious silicon obtained at 465°C by low pressure chemical vapor deposition (LPCVD) of disilane Si_2H_6, under pressure of 200 mTorr. After a standard clean, the films are deposited on oxidized monocrystalline silicon substrates (P-type, <111>, 120 nm of thermal oxyde SiO_2). They are then boron implanted with a dose of 4×10^{15} atoms/cm^2 at an energy of 15 keV [9]. In order to avoid long redistributions, post-implantation annealing was carried out at a relatively low temperature (700°C) and short times ranging between 1 and 30 minutes. The experimental doping profiles have been obtained by secondary ion mass spectrometry (SIMS).

Salah Abadli and Farida Mansour are with the department of Electronics, University Mentouri Route d'Ain EL-Bey Constantine, Constantine 25000, Algeria (e-mail: abadli.salah@gmail.com).

B. Initial Conditions

The initial total boron distribution profile (before annealing) was easily simulated by the use of an analytical Gaussian expression known by the three following parameters: the ion implantation dose Q_d, the projected range R_p, and the straggle ΔR_p, given by [10]:

$$C(x)=\frac{Q_d}{\sqrt{2\pi}\Delta R_p}\exp\left(-\frac{(x-R_p)^2}{2\Delta R_p^2}\right) \qquad (1)$$

Figure 1 shows the good superposition of simulated profile and SIMS profile before annealing. This profile will be used as reference or initial condition during theoretical simulation step.

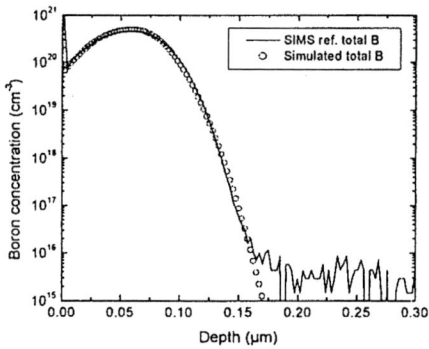

Fig. 1. Superposition of the simulated profile with the experimental SIMS profile before annealing.

Starting from the results of Solmi et al. [11], we can tell that the implantation dose of 4×10^{15} atoms/cm^2 introduce the exceeding of the solubility solid limit at these annealing temperatures. In this case, the doping excess precipitates and forms inactive and immobile clusters [4], [7]. In the reality, the state of the dopant related to this parameter was not given in a clear and single way in the literature. The ones suggests that dopant segregate to the grain boundaries [6], [12], while the others propose the possibility of clustering even at concentrations far below the solubility limit under the supersaturation of self-interstitials [7], [13].

III. DIFFUSION MODEL

Different models had been proposed for the study and the simulation of the complex mechanisms of boron diffusion and activation in single-crystal and polycrystalline silicon during the thermal post-implantation annealing [4], [9], [14]. By using the informations of each model, we describe now a more detailed model, adapted to the granular structure of polysilicon and to the effects of very strong concentrations. The diffusion is controlled by two partial differential equations (PDE's) coupled by a term representing the dopant transfer between the grains and the grain boundaries. The total concentration will be divided between the grains and the grains boundaries. The effects connected to the heavily concentrations were combined with the dopant diffusion coefficients in the grains and the grain boundaries. For a one-dimensional diffusion this model takes the following form:

$$\frac{\partial C_g}{\partial t}=\frac{\partial}{\partial x}\left(D_g^{eff}\frac{\partial C_g}{\partial x}\right)-k_t^{eff}\left(C_g-\frac{C_{gb}}{k_{seg}}\right) \qquad (2)$$

$$\frac{\partial C_{gb}}{\partial t}=\frac{\partial}{\partial x}\left(D_{gb}^{eff}\frac{\partial C_{gb}}{\partial x}+\frac{D_{gb}^{eff}}{L_g}C_{gb}\frac{\partial L_g}{\partial x}\right)+k_t^{eff}\left(C_g-\frac{C_{gb}}{k_{seg}}\right) \qquad (3)$$

C_{gb} is the dopant concentration in the grain boundaries and C_g is the dopant concentration inside the grains. The total concentration in the studied layers is thus, the sum of the two concentrations ($C_{total}=C_g+C_{gb}$). D_g^{eff} and D_{gb}^{eff} are respectively the effective diffusion coefficients in the grains and the grain boundaries. They are identified by:

$$D_g^{eff}=D_i\frac{1+\beta(p/n_i)}{1+\beta}\left(1+\frac{C_g}{\sqrt{C_g^2+4n_i^2}}\right)\left(1+m\left(\frac{C_g}{C_{sol}}\right)^{2m}\right)^{-1} \qquad (4)$$

$$D_{gb}^{eff}=D_iF_a\left(1+\frac{C_{gb}}{\sqrt{C_{gb}^2+4n_i^2}}\right)\left(1-\exp\left(\frac{-E_b}{KT}\right)\right) \qquad (5)$$

$$D_i=D_0\exp\left(\frac{-E_a}{KT}\right) \qquad (6)$$

D_i is the intrinsic diffusivity in single-crystal silicon, its value varies exponentially with the temperature T and the activation energy E_a associated with the diffusion process. In this study, we took the value of E_a=3.46eV [15]. D_0 is the diffusivity pre-exponential factor. The effective diffusion coefficient in the grains is greatly linked to the high concentration effects. It depends on the vacancies concentrations in the grains, under the effect of the ratio of the diffusivity induced by the charged vacancies on the global diffusivity induced by the neutral vacancies $\beta=D_i^+/D_i^0$ [9], [15], of holes concentration p, and of the intrinsic concentration n_i. D_g^{eff} depends also, of the doping solubility solid limit C_{sol}, as well as, of the clusters size by the maximum number m of self-interstitials sites occupied by clusters [4], [14]. Concerning the effective diffusion coefficient in the grain boundaries, it is well controlled by the trapping and the segregation to the grain boundaries [16]. These two effects are obviously related to the energy barrier E_b to the grain boundaries, it even, depends on the impurities concentration, the average size of the grains, and the traps density [17], [18]. F_a is a pre-exponential factor for the adjustment of the intrinsic diffysivity in polycrystalline silicon thin films.

The coupling between the PDE's (2) and (3) is ensured by the term representing the effective dopants transfer from the grains to the grain boundaries and vice versa. The net effective transfer rate is given by [16], [19]:

$$k_t^{eff}=\frac{D_g}{L_g}\left(\frac{4}{L_g}+\frac{1}{2\sqrt{D_gt}}\right)+\frac{2\alpha}{L_g}\frac{\partial L_g}{\partial t} \qquad (7)$$

The effective doping transfer rate between the grains and the grain boundaries depends primarily of the change of the grain average diameter L_g during thermal annealing. α is a factor of adjustment; in our work we take $\alpha=1$. The grains-growth is proportional to the square root of time [16], [20]:

$$L_g(t)=\sqrt{L_{g0}^2+2\sigma t} \qquad (8)$$

$L_g(t)$ represents the average grains size after time t and L_{g0} represents the initial average grains size after deposition of polysilicon layer. σ is a parameter which depends on the grain boundary mobility and the grain boundary energy. Its value depends on the local Fermi level. Concerning the dopant segregation, it is expressed by using the results of Mandurah et al. [21] and Swaminathan et al. [22]:

$$k_{seg}=2k_{seg0}\frac{e_{gb}}{L_g} \qquad (9)$$

k_{seg0} is the thermal equilibrium grain boundary segregation coefficient and e_{gb} is the average width of a grain boundary. Segregation coefficient is described by the expression similar to one of Mandurah et al. [21]: $K_{seg0}= K_0 exp(0.456/KT)$.

IV. RESULTS AND DISCUSSION

Simulated boron diffusion profiles were calculated by means of a realized program; which makes the numerical resolution of the differential equations (2) and (3); while using an implicit method of finite differences. From the good superposition of the simulated profiles and SIMS profiles, we can illustrate the significant role of the grains growing in size and that of the dopant trapping and segregation to the grain boundaries. The adjustment is obtained while varying the following parameters: L_{g0}, $F_a=D_{poly}/D_{mono}$, m, β, k_{seg}. Figures 2 5, show clearly the good adjustment of the simulated profiles with SIMS profiles. In these figures, we notice good matching, particularly for the profiles shoulder that occurs for the boron solubility limit. The good adjustment indicates the validation of this model and confirms that the grins-growth in size play a significant role for the precise determination of the diffusion profiles. We can tell that the boron diffusivity in the grains is not very different to that in the grain boundaries. This is justified by the significant reduction of the diffusion coefficient in the grain boundaries caused by the important trapping-segregation effects. On the other hand, diffusivity in the grains at this time is accelerated by the self-interstitials clusters formation which causes TED. The effective diffusion coefficient in the grain boundary depends much on E_b and thus on L_g. Figure 6 shows obviously the dependence of the crystallization process of the dopant concentration during annealing. It is clear that if the concentration exceeds the solid solubility limit, the grains-growth starts to tend towards a limiting. In effect, the effective diffusion coefficient in grain boundary depends much on E_b, L_g and thus on the density of traps.

The vacancies concentrations effect on boron diffusivity depends principally of β, its optimal value obtained after profiles adjustment is 0.13, which is in good agreement with that in the results of Mahamdi et al. [9] and Giroult et al. [23]. The adjustment factor of diffusivity in the grain boundaries given in the theory by the ratio D_{poly}/D_{mono} is approximately 100 in our simulation. This means that the boron diffusivity in the grain boundaries is about 100 times greater than that in the grains. This result is consistent with that obtained by Probst et al [15]. In this work, the number of self-interstitials to be trapped in the boron clusters m, takes the value of $m=2$, led to the best fitting for all the profiles. This m value is approximately similar to that obtained by Uematsu for the monocrystalline silicon [4], [7]. In consequence, the presence of high self-interstitial concentration causes transient enhanced diffusion (TED) of dopant atoms and their precipitation into clusters at a concentration well below the solubility solid limit. This was observed in rigorous manner in the simulated profiles by the shoulder (bump) of the diffusion profiles that occurs with the boron solubility solid limit. This effect was also discussed in research of Pelaz et al. [5], [13] and research of Mahamdi et al. [9].

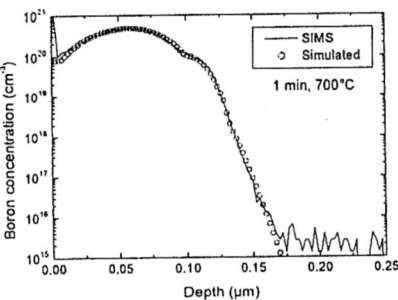

Fig. 2. Superposition of the simulated profile with the experimental SIMS profile after annealing (700°C/1min)

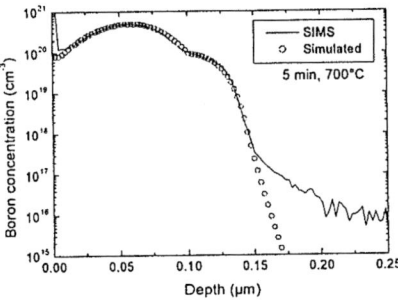

Fig. 3. Superposition of the simulated profile with the experimental SIMS profile after annealing (700°C/5min)

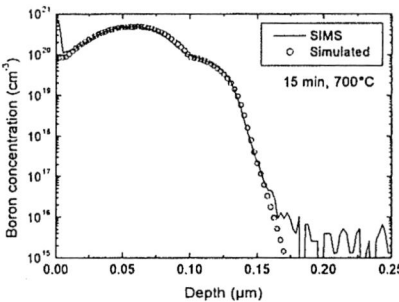

Fig. 4. Superposition of the simulated profile with the experimental SIMS profile after annealing (700°C/15min)

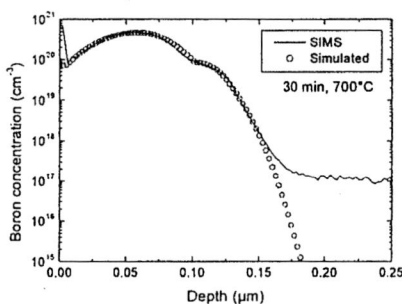

Fig. 5. Superposition of the simulated profile with the experimental SIMS profile after annealing (700°C/30min)

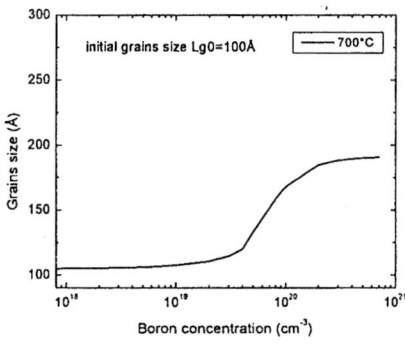

Fig. 6. Growth of grains as a function of the boron concentration during 30 minutes of annealing

V. CONCLUSIONS

Growth of grains and clustering effects are two very significant parameters for the precise simulation of the diffusion profiles. In addition, we can note that the boron transient enhanced diffusion (TED) does not depend only on the trapping and segregation to the grains boundaries but also on the clustering effect to the grains; under the effects of the strong concentrations.

REFERENCES

[1] B. Yu, D. H. Ju, W. C. Lee, N. Kepler, T. J. King and C. Hu, "Gate engineering for deep-submicron CMOS transistors," *IEEE, Trans. Electron Devices*, vol. 45, pp. 1253-1262, 1998.

[2] A. J. Walker, S. B. Herner, T. Kumar and En-H. Chen, "On the conduction mechanism in polycrystalline silicon thin-film transistors," *IEEE, Trans. Electron Devices*, vol. 51, No. 11, pp. 1856-1866, 2004.

[3] L. Pelaz, V. C. Venezia, H. J. Gossmann, G. H. Gilmer, A. T. Fiory, and C. S. Rafferty, "Activation and deactivation of implanted B in Si," *Appl. Phys., Lett.*, vol. 75, No. 5, pp. 662-664, 1999.

[4] Masashi Uematsu, " Simulation of high-concentration boron diffusion in silicon during post-implantation annealing," *Jpn. J. Appl. Phys.*, vol. 38, pp. 3433-3439, 1999.

[5] M. Jaraiz, G. H. Gilmer and J. M. Poate, "Atomistic calculations of ion implantation in Si: point defect and transient enhanced diffusion phenomena," *Appl. Phys., Lett.*, vol. 68 No. 3, pp. 409-411, 1996.

[6] H. Schaber, R. V. Criegern and I. Weitzel, " Analysis of polycrystalline silicon diffusion sources by secondary ion mass spectrometry," *J. Appl. Phys.*, vol. 58, No. 11, pp. 4036-4042, 1985.

[7] Masashi Uematsu, " Simulation of clustering and transient enhanced diffusion of boron in silicon," *J. Appl. Phys.*, vol. 84, No. 9, pp. 4781-4787, 1998.

[8] J. R. Pfiester, F. K. Baker, T. C. Mele, H. H. Tseng, P. J. Tobin, J. D. Hayden, J. W. Miller, C. D. Gunderson and L. C. Parillo, "The effects of boron penetration on P+ polysilicon gated PMOS devices," *IEEE Trans. Electron Devices*, vol. 37, No. 8, pp. 1842-1847, 1990.

[9] R. Mahamdi, F. Mansour, E. Scheid, P.T. Boyer and L. Jalabert, "Boron diffusion and activation during heat treatement in heavily doped polysilicon thin films for P+ Metal-Oxyde-Semiconductor transistors gates," *Jpn. J. Appl. Phys.*, vol. 40, pp. 6723-6727, 2001.

[10] R. W. Cahn, P. Haasen, and E. J. Kramer, *Electronic structure and properties of semiconductors*, vol. 4, April 1991, pp. 264-275.

[11] S. Solmi, F. Baruffaldi and R. Canteri, "Diffusion of boron in silicon during post-implantation annealing," *J. Appl. Phys.*, vol. 69, No. 4, pp. 2135-2142, 1991.

[12] S. Batra, M. Manning, C. Dennison, A. Sultan, S. Bhattacharya, K. Park, S. Banerjee, M. Lobo, G. Lux, C. Kirschbaum, J. Noberg, T. Smith and B. Mulvaney, "Discontinuity of B-diffusion profiles at the interface of polycrystalline Si and single crystal Si," *J. Appl. Phys.*, vol. 73, No. 8, pp. 3800-3804, 1993.

[13] L. Pelaz, G. H. Gilmer, H. J. Gossmann, C. S. Rafferty, M. Jaraiz and J. Barbella, "B cluster formation and dissolution in Si: A scenario based on atomic modeling," *Appl. Phys., Lett.*, vol. 74, pp. 3657-3660, 1999.

[14] H. Puchner and S. Selberherr, " An advanced model for dopant diffusion in polysilicon," *IEEE, Trans. Electron Devices*, vol. 42, No 10, pp. 1750-1754, 1995.

[15] V. Probst, H. J. Bohm, H. Schaber, H. Oppoler and I. Weitzel, "Analysis of polysilicon diffusion sources," *J. Electrochem Soc.*, vol. 135, No. 3, pp. 671-676, 1988.

[16] A. D. Sadovnikov, "One-dimensional modeling of high concentration boron diffusion in polysilicon-silicon structures," *Solid-State Electronics*, vol. 34, No. 9, pp. 969-975, 1991.

[17] J. Y. W. Seto, "The electrical properties of polycrystalline silicon films," *J. Appl. Phys.*, vol. 46, No. 12, pp. 5247-5254, 1975.

[18] G. Baccarani, B. Ricco and G. Spadini, "Transport properties polycrystalline silicon films," *J. Appl. Phys.*, vol. 49, pp. 5565-5570, 1978.

[19] S. K. Jones and C. Hill, "Modeling dopant diffusion in polysilicon," *Simulation of Semiconductor Devices and processes*, vol. 3, pp. 441-449, 1988.

[20] H.-J. Kim and C.V. Thompson, "Kenetic modeling of grains growth in polycrystalline silicon films doped with phosphorus or boron," *J. Electrochem. Soc.*, vol. 135, pp. 2312-2316, 1988.

[21] M. M. Mandurah, K. C. Saraswat, C. R. Helms and T. I. Kamins, "Dopant segregation in polycrystalline silicon," *J. Appl. Phys.*, vol. 51, No. 11, pp. 5755-5763, 1980.

[22] B. Swaminathan, E. Demoulin, T. W. Sigmon, R. W. Dutton and R. Rif, "Segregation of arsenic to the grain boundaries in polycrystalline silicon," *J. Electrochem Soc.*, vol. 127, No. 10, pp. 2227-2229, 1988.

[23] G. Giroult, A. Nouailhat and M. Gauneau, "Study of Wsi2 / polycrystalline silicon / monocrystalline silicon structure," *Appl. Phys.*, vol. 67, pp. 515-523, 1990.

Analysis and Modeling of RF-MEMS Disk Resonator

Mostafa M. Sakr, *Student Member, IEEE*, Mousa Khalid El-Shafie, and Hany Fikry Ragai,
Member, IEEE

Abstract— This paper presents the analysis and modeling of MEMS disk resonator. The main parameters that describe the operation of the disk are the quality factor, natural frequency and equivalent mass. In this work the frequency relation versus Poisson's ratio of the material and a general equation for the equivalent mass of disk resonator are obtained for different vibration modes of the disk resonator. The resulting model is formulated such that it can be coded by any suitable language such as systemC, verilog-AMS or VHDL-AMS. The model is verified by ANSYS.

GLOSSARY

ρ	density $\left(Kg \cdot m^{-3} \right)$
ω	angular frequency $\left(rad^{-1} \right)$
ν	Poisson's ratio $(-)$
E	Young's modulus $\left(kg \cdot m^{-1} \cdot s^{-2} \right)$
R	disk radius (m)
K	dimensionless frequency parameter $(-)$
t_{disk}	disk thickness (m)
$V_n(r)$	velocity of any point $\left(m \cdot s^{-1} \right)$
$V_n(R)$	velocity at the edge of the disk $\left(m \cdot s^{-1} \right)$
m_{re}	resonator effective mass (Kg)
k_r	resonator effective stiffness $\left(kg \cdot s^{-2} \right)$
c_r	resonator effective damping $\left(kg \cdot s^{-1} \right)$
Q	resonator quality factor $(-)$
$\left(\dfrac{B}{A} \right)$	ratio between the radial component and the tangential component
λ' and μ	Lamé constants $\left(kg \cdot m^{-1} \cdot s^{-2} \right)$
$\Gamma(n)$	Gamma function $= (n-1)!$
$_pF_q$	generalized hypergeometric function
$_p\tilde{F}_q$	regularized generalized hypergeometric function

I. INTRODUCTION

DISK resonators are emerging as a potential candidate for on-chip small percent bandwidth filters [1] and high Q reference oscillators [2]. The operating frequencies of disk resonators reach UHF band [3] and as dimensions scales down operating frequencies increases. To integrate disk resonators with existing circuit or system design flow accurate and flexible models are required.

Disk resonators are distributed-parameter systems as the wave length of the disk vibration is comparable to the disk dimensions. Any distributed-parameter system exhibits infinite degrees of freedom (i.e. infinite mechanical resonances or modes [4].) Each mode is represented by an equivalent lumped-parameter mass-spring system as shown in figure 1.

In this model the effective stiffness is given by

$$k_r = \omega_n^2 m_{re} \tag{1}$$

and damping element value is given by

$$c_r = \frac{\sqrt{k_r m_{re}}}{Q} \tag{2}$$

to describe each mode three parameters are required: quality factor (Q), natural frequency (ω_n), and equivalent mass (m_{re}) for the mode. The quality factor is to be measured experimentally. Natural frequency which results from the solution of the frequency equation and determination of the equivalent mass will be discussed in this paper.

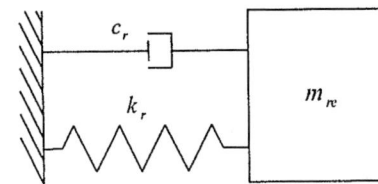

Fig. 1. Equivalent lumped-parameter mass-spring system

II. SOLUTION OF THE WAVE EQUATION

The following results are based on some assumptions [5].

1. The amplitude of vibration and the stress-strain relation is linear.

2. There are no internal losses and the vibration occurs in vacuum.

3. There is no body forces such as gravity or magnetic fields applied to the resonator.

TABLE I
SUMMARY OF RADIAL DISPLACEMENT AND THE FREQUENCY EQUATIONS

Mode	Radial $n = 0$	Compound $n > 0$
U_n, radial displacement	$A h J_1(hr)$	$\left[A\left(\dfrac{\partial}{\partial r}\right) h J_n(hr) + nB\left(\dfrac{1}{r}\right) J_n(kr) \right]\cos n\theta$
Frequency equation	$\dfrac{(\zeta/\xi) J_0(\zeta/\xi)}{J_1(\zeta/\xi)} = 1 - \upsilon$	$\left[\dfrac{(\zeta/\xi) J_{n-1}(\zeta/\xi)}{J_n(\zeta/\xi)} - n - Q \right]\left[\dfrac{\zeta J_{n-1}(\zeta)}{J_n(\zeta)} - n - Q \right] = (nQ - n)^2 , n > 1$ $Q = \zeta^2/(2n^2 - 2)$

The elastic wave equations for disk resonators were solved by Love [6], and Onoe [7]. From their solution the general frequency equation and the general displacement relations are derived.

There are two cases $n = 0$ and $n > 0$. In case of $n = 0$ there are no nodal diameters, but there is circular nodal lines. Table 1 summarizes the radial displacement and the frequency equations for disk resonator. Where,

$$\xi^2 = \frac{2}{(1-\upsilon)} , \ \zeta = kR$$

$$, h^2 = \frac{\rho\omega^2}{(\lambda' + 2\mu)} , \ k^2 = \frac{\rho\omega^2}{\mu}$$

$$\lambda' = \frac{\nu E}{(1 - \nu^2)} , \ \mu = \frac{E}{2(1+\nu)}$$

Only radial displacement is considered as a result to the excitation technique used. Excitation is done mainly thought the variation of the gap in radial direction as the electric field lines will always terminate perpendicular to the surface of the disk, and the electrostatic force will be in the direction of the field lines. These forces will mainly excite radial displacement. It worth noting that the radial displacement in the compound modes ($n > 0$) includes a tangential component. For $n = 0$ modes there are two independent modes: one is characterized by the absence of the tangential components (radial mode), and the other is characterized by the absence of the radial component (tangential mode) [7]. Pure tangential modes cannot be excited.

Also, Onoe [7] got the ratio between the radial component and the tangential component $\left(\dfrac{B}{A}\right)$ as,

$$\frac{B}{A} = \frac{2n\left((\zeta/\xi)\dfrac{J_{n-1}(\zeta/\xi)}{J_n(\zeta/\xi)} - (n+1) \right)}{2\zeta\dfrac{J_{n-1}(\zeta)}{J_n(\zeta)} + \zeta^2 - 2n(n+1)} \tag{3}$$

III. NATURAL FREQUENCY

Natural frequency in defined by [7]

$$f = \frac{K}{2R}\sqrt{\frac{E}{\rho}} \tag{4}$$

K for each mode is expressed by [6]

$$K = \frac{(\zeta/\xi)}{\pi\sqrt{(1 - v^2)}} = \frac{\zeta}{\pi\sqrt{(2 + 2v)}} \tag{5}$$

where (ζ/ξ) and ζ are the roots of the frequency equations given in table 1. To solve these equations a numerical solution software package is required, for this work MATLAB is used to determine the dimensionless frequency parameter K for: Radial mode (fundamental, 1st, 2nd, 3rd, etc. harmonics), Compound modes ($n = 2$, $n = 4$, $n = 6$) with their harmonics. The results of K versus v is fitted in a forth order polynomial of the form

$$K = p_1 \times v^4 + p_2 \times v^3 + p_3 \times v^2 + p_4 \times v + p_5 \tag{6}$$

where the constants p_1, p_2, p_3, p_4, p_5 are given in table 2 for different mode orders and their harmonics.

For compound modes even modes were chosen as they provide symmetrical mode shape.

IV. EQUIVALENT MASS

The equivalent mass of a disk is defined as the equivalent lumped mass equivalent to the distributed parameter disk, at a point in a given direction and in the region of the natural frequency [5]. To get an equivalent mechanical lumped elements model we need to calculate the equivalent mass of certain mode. That is done by equating the total kinetic energy of the disk at resonance by one-half the squared velocity at a certain point and direction multiplied by the equivalent mass required.

$$\frac{1}{2} M_{eq} V_n^2 \Big|_{r=R} = \frac{1}{2}\int_0^{t_{disk}} \int_0^{2\pi} \int_0^R V_n^2(r)\, r\, dr\, d\theta\, dz \tag{7}$$

then,

$$M_{eq} = \frac{\pi \rho t_{disk} \int_0^R r V_n^2(r)\, dr}{V_n^2(R)} \tag{8}$$

and

$$V(r,t) = \frac{dU(r,t)}{dt} \tag{9}$$

Noting that the time dependence of $U(r,t)$ is $e^{-j\omega_n t}$ then to calculate M_{eq} let $V(r,t) = V_o \times U_n e^{-j\omega_n t}$, then the mass equation becomes

$$M_{eq} = \frac{\pi \rho t_{disk} \int_0^R r U_n^2(r) dr}{U_n^2(R)} \quad (10)$$

and the mass is calculated by calculating the integration $\int_0^R r U_n^2(r) dr$ for $n = 0$ and for $n > 0$.

A. For $n = 0$

$$M_{eq} = \frac{1}{2} \pi \rho t_{disk} R^2 \left(1 - \frac{J_0(hR) J_2(hR)}{J_1^2(hR)} \right) \quad (11)$$

B. For $n > 0$

From Onoe's equation for $n > 0$,

$$U_n = \begin{bmatrix} A\left(\dfrac{\partial}{\partial r}\right) h J_n(hr) \\ + nB\left(\dfrac{1}{r}\right) J_n(kr) \end{bmatrix} \cos n\theta \quad (12)$$

, which can written as

$$U_n = \cos n\theta \times U_r(r) \quad (13)$$

substitute in (9)

$$M_{eq} = \frac{\pi \rho t_{disk} \int_0^R r U_r^2(r) dr}{U_r^2(R)} \quad (14)$$

the integral $\int_0^R r U_n^2(r) dr$ is solved in (14). the integration of this equations give the following closed forms, while two of the integrals does not have a closed form solution and will be computed numerically for each mode of the supported modes.

V. VERIFICATION

The results obtained from the analytical equations were compared to the values extracted from ANSYS. There was a slight difference between the simulated and calculated values. These differences are not critical as the proposed model is to give the designer rough values to the disk parameters that satisfy the required performance.

Figures 2 and 3 show the calculated frequency parameter versus Poisson's ratio for n=0 and 4 respectively.

VI. CONCLUSION

Equations (11) and (15) along with equation (6) describes the disk resonator operation. Equation (6) is verified, and it provides a good approximation for the relation between Poisson's ratio and the frequency.

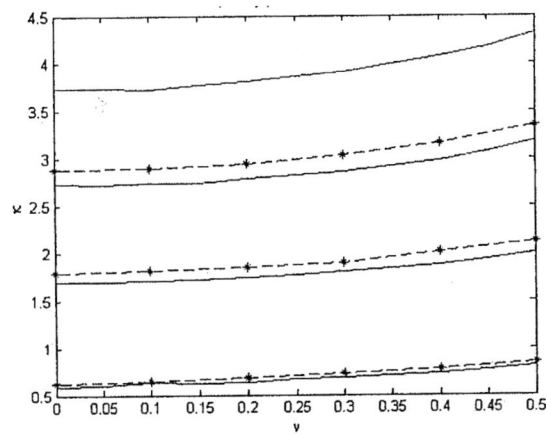

Fig. 2 Dimensionless frequency parameter versus Poisson's ratio for n=0
Solid line: calculated data, dashed lines: ANSYS data

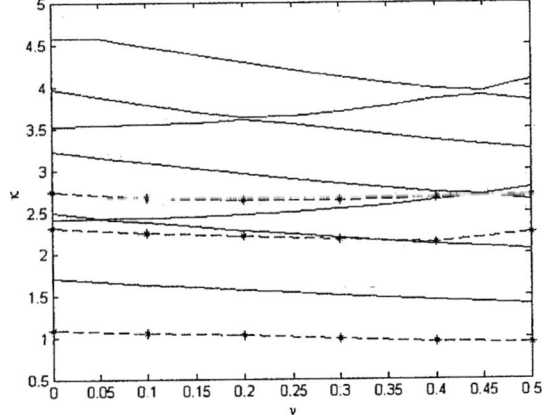

Fig. 3 Dimensionless frequency parameter versus Poisson's ratio for n=4
Solid line: calculated data, dashed lines: ANSYS data

REFERENCES

[1] Sheng-Shian Li, Yu-Wei Lin, Zeying Ren, and Clarck T.-C. Nguyen, "Disk-array design for suppression of unwanted modes in micromechanical composite-array filter," Proceedings, IEEE MEMS conference, Istanbul, turkey, 2006, pp.866-869

[2] Brian P. Otis, and Jan M. Rabaey, "A 300-μW 1.9-GHz CMOS oscillator utilizing micromachined resonators," IEEE Journal of Solid-State Circuits, vol. 38, July 2003, pp. 12711274

[3] John R. Clarck, Wan-Thai Hsu, Mohamed A. Abdelmoneum, and Clarck T.-C. Nguyen, "High-Q UHF Micromechanical radial-contour mode disk resonators," IEEE Journal of Micromechanical systems, vol. 14, December 2005, pp. 1298-1310

[4] Harrie A. C. Tilmans, "Equivalent circuit representation of electromechanical transducers: I. Lumped-parameter systems," Journal of Micromechanics and Microengineering, ,vol. 6, pp. 157-176,1996

[5] Robert A. Jonson, "Mechanical Filters in Electronics," John Wiley and Sons, 1983

[6] A. E. H. Love, "A Treatise on The Mathematical Theory of Elasticity," Dover publication, 1944

[7] Morio Onoe, "Contour Vibrations of Isotropic Circular Plates," Journal of The Acoustical Society of America, Vol. 28, No. 6, Nov. 1956, pp. 1158-1162

$$\int_0^R rU_r^{\;2}(r)dr = \left[\left(\frac{B}{A} \right)_n \times n \right]^2$$

$$\times \frac{2^{-2n-1}(kR)^{2n}\,_2F_3\left(n, n+\frac{1}{2}; n+1, n+1, 2n+1; -k^2R^2 \right)}{n\Gamma(n+1)^2}$$

$$-2\times \left(\frac{B}{A} \right)_n \times n^2 \times \int_0^R \frac{1}{r} \times J_n(hr) \times J_n(kr)dr$$

$$+2\times h \times \left(\frac{B}{A} \right)_n \times n \times \int_0^R J_{n-1}(hr) \times J_n(kr)dr$$

$$+n^2 \times \frac{1}{2}R^2 \left(J_n^2(hR) - J_{n-1}(hR)J_{n+1}(hR) \right)$$

$$+\frac{3h}{4} \times \frac{1}{2}R^2 \left(J_{n-1}^2(hR) - J_{n-2}(hR)J_n(hR) \right)$$

$$-2 \times n \times h \times \frac{2^{-2n}R(hR)^{2n-1}\,_2F_3\left(n, n+\frac{1}{2}; 2n, n+1, n+1; -h^2R^2 \right)}{n\Gamma(n)\Gamma(n+1)}$$

$$+\frac{h}{4} \times 2^{-n}R(hR)^n\Gamma\left(\frac{n+1}{2} \right)\,_1\tilde{F}_2\left(\frac{n+1}{2}; n, \frac{n+3}{2}; -\frac{1}{4}h^2R^2 \right)$$

$$\text{Where, } \,_p\tilde{F}_q = \frac{\,_pF_q(a;b;z)}{\left(\Gamma(b_1)\dots\Gamma(b_q) \right)}$$

(15)

TABLE II
VALUE OF CONSTANTS IN EQUATION (6) FOR DIFFERENT MODE ORDERS AND THEIR HARMONICS

Mode order n	Harmonic number	p_1	p_2	p_3	p_4	p_5
(radial)	1st	-4.9619	6.342	-2.0371	0.49166	0.58498
	2nd	1.5972	-0.55632	0.96762	0.058959	1.696
	3rd	14.047	-11.879	4.8915	-0.32127	2.7242
	4th	17.671	-16.621	7.39	-0.59412	3.7367
	5th	4.4232	-1.3613	2.4651	0.050428	4.7282
	6th	0.073949	4.2789	1.5417	0.27979	6.7265
2	1st	0.1684	0.14323	-0.60516	0.0016954	1.1559
	2nd	27.558	-28.508	6.9266	0.3355	1.6375
	3rd	-25.693	28.393	-8.0312	0.47853	1.8918
	4th	0.38114	0.32393	-1.3692	0.0038337	2.6154
	5th	-6.7112	-3.909	2.8564	0.95394	2.6843
	6th	9.8042	3.7456	-4.7849	0.38078	3.3252
4	1st	0.24883	0.21164	-0.89418	0.0025115	1.708
	2nd	-17.201	21.255	-9.8881	1.3926	2.4192
	3rd	-19.74	10.828	0.35781	0.42904	2.4813
	4th	40.138	-31.906	6.3704	-0.6147	3.2403
	5th	46.833	-42.674	6.6114	1.5793	3.519
	6th	-100.52	88.395	-20.171	1.0198	3.9609
6	1st	0.32581	0.27713	-1.1708	0.0032897	2.2364
	2nd	0.44586	0.37869	-1.6012	0.0044673	3.0586
	3rd	23.824	-33.276	10.362	0.64463	3.1717
	4th	-20.183	33.065	-12.582	0.93048	3.8183
	5th	-26.88	37.938	-20.479	3.6586	4.3189
	6th	37.152	-62.305	27.303	-2.4358	4.5907

Behavioral Modeling of RF-MEMS Disk Resonator

Mousa Khalid El-Shafie, Mostafa M. Sakr, *Student Member, IEEE*, and Hany Fikry Ragai, *Member, IEEE* Ain Shams University, Electronics & Comm. Department, Cairo, Egypt.

Abstract — This paper presents Behaviora l modeling of RF - MEMS disk resonator through driving its equivalent circuit model. The model is operating in the first two radial contour modes and the results were verified with ANSYS . A Pierce oscillator utilizing the resonator model is designed to demonstrate the performance of the model on the oscillator. The oscillator performance like frequency and phase noise is presented.

I. INTRODUCTION

Over the other resonator structures (e.g. comb drives , beam resonators) the disk resonators have many advantages : a. Disks operating in radial modes exhibit high stiffness , which give the resonator high kinetic energy making it less susceptible to losses, while retaining high Q, b. Large spring constants (stiffness) allows high frequencies (beyond GHz), while maintaini ng large masses , c. Large mass means large surface-to-volume ratio , which gives better coupling coefficients [1].

To enable system level simulation and architecture exploration there is a need for a unified design environment to represent different energy domain such as electrical and mechanical. VHDL-AMS is a good candidate for such multi-domain systems.

To benefit form VHDL -AMS potential flexible and accurate models must be available . These models must model the effect of design parameters and process p arameters on the performance of disk resonators.

II. DISK RESONATOR OPERATION

Figure 1 presents a perspective schematic view of the disk μ-resonator identifying key dimensions and indicating a preferred (2-port) bias and excitation scheme . The resonator consists of a disk suspended above the substrate with a single anchor at its center . Plated metal input electrodes surround the perimeter of the disk , separated by a narrow air (or vacuum) gap that defines the capacitive , electromechanical transducer of this device. To operate the device , a DC bias voltage V_P is applied to the structure , while an AC input signal is applied to the input electrode , resulting in a time varying electrostatic

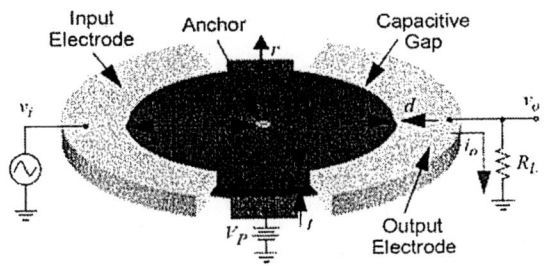

Fig. 1 Disk resonator structure [2]

force acting radially on the dis k. When the input signal , and hence the force that it generates , are acting at the resonant frequency of the device , that force is effectively multiplied by the Q of the resonator.

III. PROPOSED MODEL

A. Model Attributes

In this section, a description of the Disk-Resonator model is introduced. Each attribute is presented, and its effect on the model is also briefly discussed.

1) Design parameters

The main design parameters that affect the operation of a disk resonator are : radius of disk , electrode-disk gap thickness, and the material used for disk . These design parameters will be the generics for the model.

2) Inputs & Outputs

Disk resonator has two types of inputs electrical inputs applied on the electrodes, mechanical input such as force that is applied directly on the disk in the case of using mechanical coupling elements to control transfer function in filter application [3].Outputs of disk resonator are the same as its inputs. It has also electrical and mechanical terminals.

3) Supported modes

Disk resonator i s a distributed parameters system as the wave length of disk vibration wave is comparable to the disk dimensions. These distributed parameters systems have an infinite modes of vibration . Each mode need to be modeled as one lumped parameters model [4].

1-4244-0899-7/06/$20.00 ©2006 IEEE.

4) Effect of DC-bias

DC-bias of the disk affects the motional resistance , the equivalent capacitance of mode, and the equivalent stiffness. Also, it affects the natural frequency of the structure.

5) Secondary effects

There are other parameters that affect the operation of the disk resonator , such as : different noise mechanisms , thickness of the disk , supporting elements, imperfections in disk material.

B. Equivalent circuit

Once the resonant frequency has been determined , other parameters such as equivalent cir cuit elements may be extracted to aid in system design . By the analogy between the lumped element equivalent mechanical model and the RLC system we can get the equivalent circuit , firstly as shown in fig . (2-a) the resonator may be represented by a lumped-element equivalent mechanical model . This model, composed of a rigid -body mass connected to a stationary base through spring and damper elements , aids in the design of more complex systems such as filters. The formulas of the effective mass m_r and the reso nance frequency as presented in [5].

$$f_o = \frac{\kappa\alpha}{R}\sqrt{\frac{E}{\rho}}, \qquad (1)$$

$$m_r = \frac{\rho\pi t}{4hJ_1(hR)}\int_0^R hJ_1(hr)\delta r, \qquad (2)$$

where, $$h = \sqrt{\frac{\omega_o^2\rho}{\dfrac{2E}{2+2\sigma}+\dfrac{E\sigma}{1-\sigma^2}}}, \qquad (3)$$

in turn the effective stiffness k_r of the resonator is given by,

$$k_r = \omega_o.m_r \qquad (4)$$

and finally, the value of the damping element is related to both m_r and k_r, as well as the Q of the resonator, by

$$c_r = \frac{\sqrt{k_r.m_r}}{Q} \qquad (5)$$

In most cases, the Q of the resonator is difficult to predict theoretically, so the damping element is typically determined empirically. The mechanical model represented can be transformed to an equivalent circuit which presented by RLC system due to the analogy between the differential that describe the two [4], where equivalent circuit related to the disk resonator parameters as follows,

$$L_x = \frac{m_r}{\eta^2} \quad , \quad C_x = \frac{\eta^2}{k_r} \quad , \quad R_x = \frac{c_r}{\eta^2}$$

$$\eta = V_p\frac{dC_o}{dx} \quad , \quad C_o = \frac{2\pi R\in_o t}{d}$$

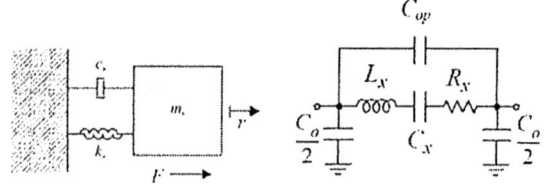

Fig. 2: Equivalent models a) Lumped element equivalent mechanical model. b) Equivalent circuit-two port setup.

Where ή is the transducer coupling coefficient at each port, V_p is dc bias voltage applied in order to linearize the relation between the force and voltage , C_o is the static capacitance between the electrodes and the disk , C_{op} represents the feed -through current in the substrate . The two port setup is used to represent the resonator to overcome the problems in the one port setup posed by shunting C_o with the system [6], This done by applying input signal on one electrode and get the output from the other el ectrode and applying the dc bias voltage directly on the structure [see figure 1]

C. VHDL-AMS Model.

Using the obtained equivalent circuit for the resonator the next step is to develop a VHDL -AMS model its generics are the resonator design parameters in ord er to enable the designer to investigate the resonator behavior with the rest of the circuit elements using a powerful circuit simulator before covering all the design steps.

For the testing proposal of the model the setup shown in fig.(3) is used, the inputs for the model are the main design parameters that affect the mechanical vibrations of the disk which are its radius and the properties of the used material . The Resonator model proposed here would be made of polysilicon with material properties Young 's modulus (150 GPa), Poisson's ratio (0.29), and density (2330 Kg/m3). The radius of the mechanical disk resonator is equal to 45 µm which is designed for a resonance frequency of 60.6 MHz, Such a frequency is chosen to allow a comparison between previous work [6]. A disk thickness of 3 µm, an air gap of 0.1 µm and a DC bias voltage of 35 volts are designed to achieve a series resistance of 1.5kΩ suitable for oscillation . A quality factor of 50,000 is assumed [7].

The simulation result of the AC analysis is compared in fig . 4 with that obtained from ANSYS to verify the model performance.

The 2006 International Conference on MEMS, NANO and Smart Systems

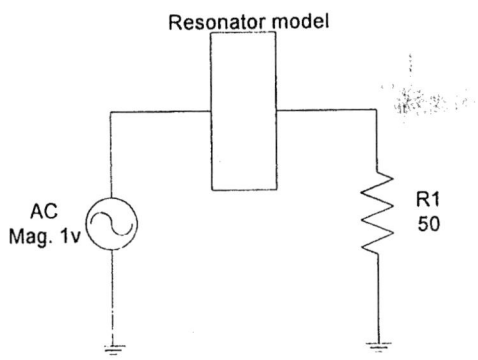

Fig. 3: Testing scheme used to measure the performance of the resonator model.

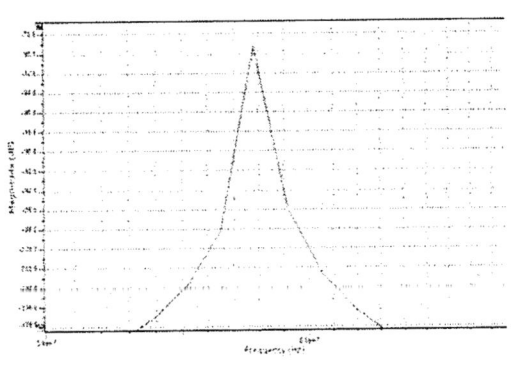

(a)

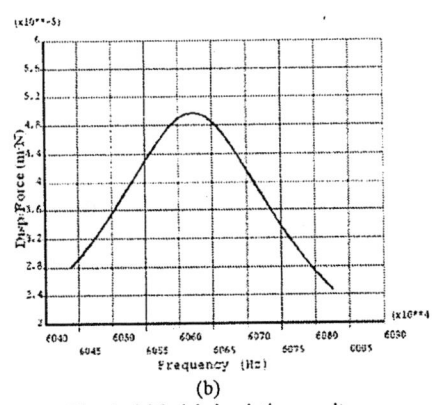

(b)

Fig. 4: a) Model simulation result.
b) ANSYS simulation result

IV. OSCILLATOR

As a demonstration for the resonator model we design a reference PIERCE oscillator utilizing an RF -MEMS resonator instead of the bulky crystal . The proposed model used as the resonator to verify its perf ormance when integrated with the other circuit components.
The oscillator design is carried out using AMS 0.35 μm technology. The role of the resistance R_{bias} is to give bias

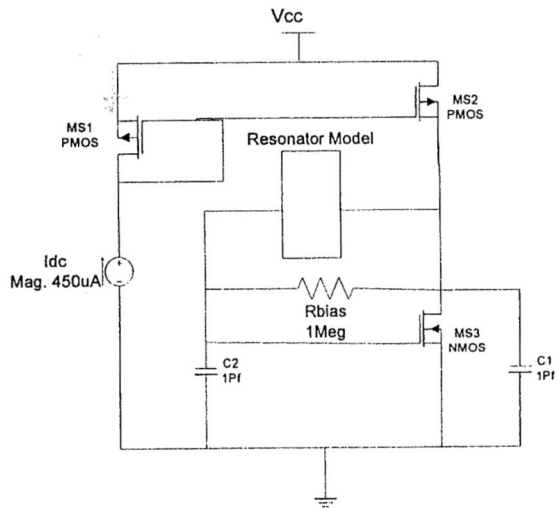

Figure 5: A PIERCE oscillator utilize disk resonator model.

to the gate since the resonator is open circuit in DC due to series resonance . This resistance , however, is chosen to be 1MΩ in order to avoid loading the resonator and degrading its quality factor . The values for C_1 and C_2 can be much smaller than those typically used in a crystal oscillator because there are no restrictions imposed by the MEMS resonator, and are assumed to be 1pF. This makes their integration on-chip much simpler. The output signal is taken at the gate and not at the drain to allow a larger signal head room and avoid distortion.

Table (1): Design summary

2 PMOS Transistors	W/L = 125μ/0.5μ
1NMOS Transistors	W/L=8μ/0.5μ
Vcc	3.3 V.
Idc	450μA
Rbias	1Meg
C1,C2	1pF

V. SIMULATION RESULTS

A. Transient Analysis:

The results of the transient analysis simulations are shown in Fig. 6 the oscillator shows a startup time of 0.5 msec and a stability time of 2 msec. This is roughly equivalent to 115,200 periods. This long startup is due to the high quality factor.

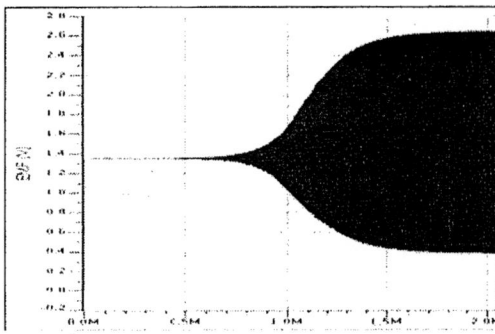

Fig. 6: Transient analysis result.

B. Phase noise:

The results of the phase noise simulation are shown in fig. 7 the designed oscillator achieves -145 dBc/Hz @ 3 KHz offset. This due to the fact that only three active elements were be used in the circuit. In addition, the very high quality factor of our resonator contributes in the reduction of the phase noise.

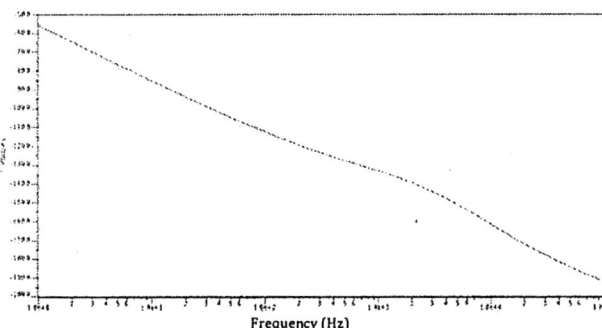

Frequency (Hz)
Fig. 7: Output phase noise at the gate

VI. CONCLUSION

In this work behavioral modeling of disk resonator was presented, through driving the equivalent circuit then using VHDL-AMS language a generic model was developed depending on the design parameter s and the model was verified using ANSYS. The unified design environment was demonstrated by the design and simulation of MEMS based Pierce oscillator operating at 60 MHz.

VII. REFERENCES

[1] C. T.-C . Nguyen, "Vibrating R F M E M S overview: applications to w ireless communications," Proceedings, Photonics West: MOEMSMEMS 2005, San Jose, California, Jan. 22-27, 2005, Paper No. 5715-201

[2] J. R. Clark, M. A. Abdelmoneum, C. T.-C. Nguyen, "U H F high - order radial-contour m ode disk resonators," Proceedings, 2003 IE E E Int. Frequency Control Symposium, Tampa, Florida, May 5-8, 2003, pp. 802-809.

[3] Y.-W. Lin, S.-S. Li, Z. Ren, and C. T.-C . Nguyen, "Low phase noise array-composite micromechanical wine-glass disk oscillator," Technical Digest, IEEE Int. Electron Devices Mtg., Washington, DC, Dec. 5-7, 2005, pp. 287-290

[4] Tilmans, H . A . C . 1996, "Equivalent Circuit Representation of Electromechanical Transducers-part I, II: lumped-parameter system s " . J. Micromech.Microeng. 6 157-76.

[5] "Analysis and Modeling of RF-MEMS Disk Resonator", Mostafa sakr, student member ieee, Mousa Khalid El-shafie, Hani fikry ragai, member ieee.

[6] J. R. Clark, W.-T. Hsu, and C. T.-C . Nguyen, "Measurement techniques for capacitively-transduced VHF-to-UHF micromechanical resonators" Digest of Technical Papers, the 11th Int. Conf. on Solid-State Sensors & Actuators (T ransducers '01), Munich, Germany, June 10-14, 2001, pp. 1118-1121.

[7] " A M E M S Disk Resonator-Based Oscillator" Moustafa M . E l Khouly, Yasseen Nada, Emad Hegazi, Hani F. Ragai, and Moustafa Y. Ghannam

Determining the Required Pulses for Controlling the Operation of Electrostatic MEMS Converters

Marwa S. Salem*, Mona S. Salem*, A. A. Zekry*, H. F. Ragai*
*Ain Shams Univ., Fac. of Eng., ECE Dept.
mona_marwa2002@yahoo.com

Abstract. – The main objective of this paper is to determine the required controlling pulses which control the charging and the discharging operations of the electrostatic MEMS converter. These converters found in the energy scavenging system for wireless sensor nodes. To achieve this objective, a SPICE model for the converter is implemented in order to describe its behavior. Thus the required controlling pulses are determined by simulating the interaction between the converter model and the energy scavenging system.

I. INTRODUCTION

An increasing focus in the researches on small wireless electronic devices has been seen in the past few years. Advances in low power Very Large Scale Integration (VLSI) design along with the low duty cycles of wireless sensors have reduced power requirements to the range of tens to hundreds of microwatts. Such low power dissipation opens up the way to power sensor nodes by scavenging energy from the environment [1, 2]. This method has been called "energy scavenging", because the node is scavenging or harvesting unused ambient energy. There are many energy scavenging technologies such as solar energy, acoustic noise, passive human power and vibration. Both solar power and vibration based energy scavenging look promising as methods to scavenge power from the environment. Solar cells are a mature technology, and one that has been profitably implemented many times in the past. Therefore the main focus of the research and development effort has been concentrated on vibration sources and vibration based electrostatic MEMS converters [3]. For scavenging the energy from vibration, an electrostatic MEMS converter is used to convert the input mechanical vibration signal to electrical signal. With advances in micro-electromechanical (MEMS) technology, it is possible to implement a self-powered system with the electrostatic MEMS converter, with conversion governed by employing low power digital control techniques. In this paper a SPICE model for an electrostatic MEMS converter which represents its behavior will be presented. By studying the interaction between the converter model and the energy scavenging system, the

required controlling pulses which control the converter operation will be determined

II. SPICE MODEL FOR THE ELECTROSTATIC MEMS CONVERTER

In this section, the elements of the SPICE model which explains the behavior of the electrostatic MEMS converter will be introduced. Figure 1 represents the input displacement (X) to the converter caused by the input mechanical vibration. It also shows the capacitances variation of the electrostatic MEMS converter with respect to time $(C-t$ curve).

In the positive half cycle of the input displacement, the converter capacitance increases from its minimum value (C^*_{min}) to its maximum value (C^*_{max}) when the input displacement increases. The converter capacitance decreases when the displacement decreases.
The converter repeats this operation in the negative half cycle of the input displacement.

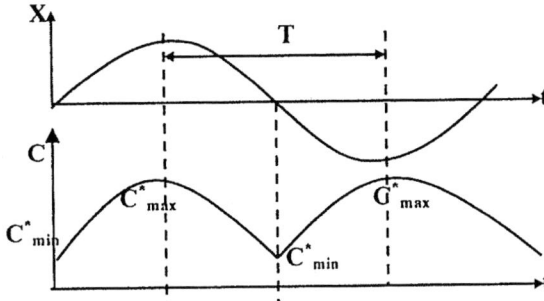

Figure 1: $C-t$ curve of the converter with input displacement

From figure 1, on can notice that the change of the converter capacitance between C^*_{max} and C^*_{min} does not depend on the path between them. Thus it is sufficient to take only two values of the converter capacitances $(C^*_{max}$ and $C^*_{min})$ in the converter model. Figure 2 represents the SPICE model for the electrostatic MEMS converter. The values of C^*_{max} and C^*_{min} depend on the converter design [4]. The coil (L1) is a model element. It is used as a charge pumping technique to transfer the charge from C^*_{max} to C^*_{min} through the switches S4 and S5. A current source (I1) is added to simulate the work done by the converter when changing its capacitances from C^*_{max} to

C^*_{min}. Thus the converter voltage increases from minimum (V_{min}) to maximum value (V_{max}) [5].

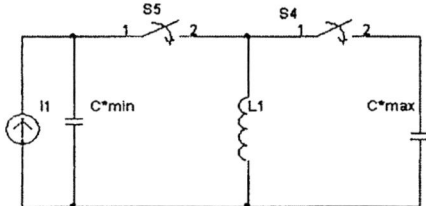

Fig.2: SPICE model for electrostatic MEMS converter

III. SYSTEM BLOCK DIAGRAM

Figure 3 represents the block diagram of the energy scavenging system. It is used to determine the required controlling pulses of the converter.

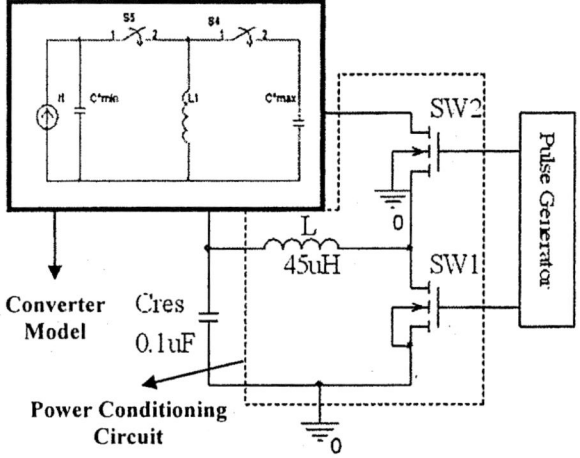

Fig.3: Energy scavenging system block diagram

C_{res} is a reservoir capacitor. It gives the initial charge for the converter which is needed to set the initial voltage on the converter to V_{min} at each beginning of the input vibration cycle [5]. It is considered as a storage element and the supply of the system. It gains energy from the system at the end of each input vibration cycle. This energy caused by the work done by the converter when converting the input mechanical vibration signal to electrical signal. The coil (L) and the switches (SW1 and SW2) represent the elements of the power conditioning circuit. It is used to condition the initial charge to be suitable for charging the converter. In addition it conditions the energy, which is given by the converter at the end of the vibration cycle, to be delivered to the load. The pulse generator is used to generate the controlling pulses. These pulses, control the on / off state of switches SW1, SW2. By using the pulse generator one can determine the specifications and the requirements of the controlling pulses which control the converter operation.

IV. DETERMINING THE REQUIRED CONTROLLING PULSES OF THE CONVERTER OPERATION

In this section, the interaction between the elements of the energy scavenging system and the converter model will be simulated. Thus the specifications of the required controlling pulses for the converter operation will be determined.

A. Charging L from C_{res}

Referring to figure 3, at the beginning of input vibration cycle, a controlling pulse from the pulse generator is applied to switch SW1 on. Thus L is charging from C_{res} to a maximum current. This current is sufficient to charge the converter at C^*_{max} by V_{min}. Thus the first charging pulse must be generated from the controller at C^*_{max}. In addition its pulse width equals to the pulse width of the controlling pulse on SW1. Figure 4 shows the simulation result of charging L from C_{res}. It shows that the reservoir voltage (V_{Cres}) is decreased. This indicates that C_{res} gives initial charge to the system at the beginning of input vibration cycle.

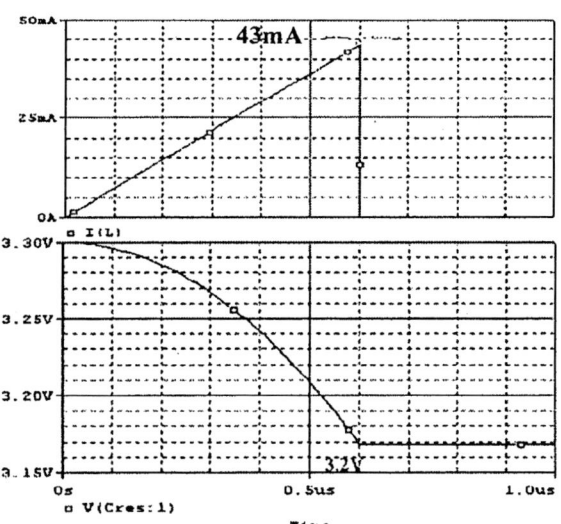

Fig.4: Charging L from C_{res}

B. Charging C^*_{max} from L

To charge C^*_{max} to V_{min} from L, L must completely discharge in C^*_{max}. Thus the coil voltage (V_L) must reach zero volts. Figure 5 shows that V_L reaches zero after one – quarter of the resonant period of the LC^*_{max} resonant circuit.

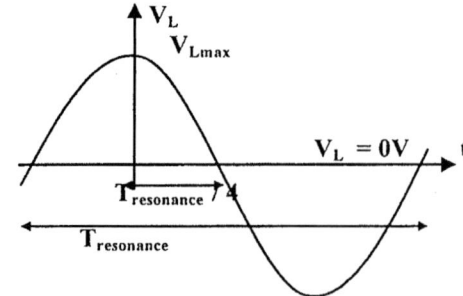

Fig.5: Coil voltage (V_L) waveform

Thus a controlling pulse is generated from the pulse generator to switch SW2 on in order to charge C^*_{max} to V_{min} from L. Therefore the second charging pulse must be generated at C^*_{max} and its pulse width equals to one quarter of the resonant period of the LC^*_{max} resonant circuit. Figure 6 shows the simulation result of charging C^*_{max} from L. It indicates that V_L reaches zero and C^*_{max} charges to V_{min}.

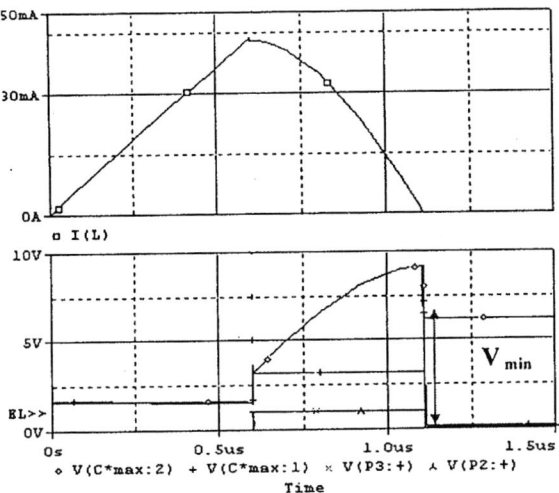

Fig.6: Charge C^*_{max} from L

L1, S4 and S5, of the converter model, are used to complete the charge transfer from C^*_{max} to C^*_{min}.

From the simulation results of transferring the charge from C^*_{max} to C^*_{min} [4] the voltage across C^*_{min} is found to be less than V_{max}. Therefore the current source (I1) is used to complete the charging of C^*_{min} to V_{max}. Thus the charging operation of the converter is completed. By this way, the SPICE model simulates the effect of the electrostatic MEMS converter when changes its capacitances C^*_{max} to C^*_{min}. Also the model simulates the increasing of the converter voltage from V_{min} to V_{max}. Thus the model represents the effect of the converter in converting the input vibration signal to electrical signal.

C. Charging L from C^*_{min}

After charging the converter to V_{max}, the converter discharges in L. So the pulse width of the first discharging pulse is determine by the quarter of the resonant period of the LC^*_{min} resonant circuit. Also the first discharging pulse must be generated when the converter capacitance at C^*_{min}.

D. Return the Charge Back to C_{res}

At the end of the vibration cycle the charge returns back to C_{res} from L. Figure 7 shows the simulation results of the coil current (i_L) and V_{Cres} over a multiple cycles of the input displacement (X). From this figure one can notice that, V_{Cres} always decreases at the beginning of the vibration cycle (due to the delivered energy from it to the system). It means that the system doesn't need to be charged from an external source at the beginning of the input vibration cycle. In addition V_{Cres}

increases at the end of the vibration cycle by a constant step. This increasing is due to the effect of the added energy caused by the work done by the converter in the conversion operation.

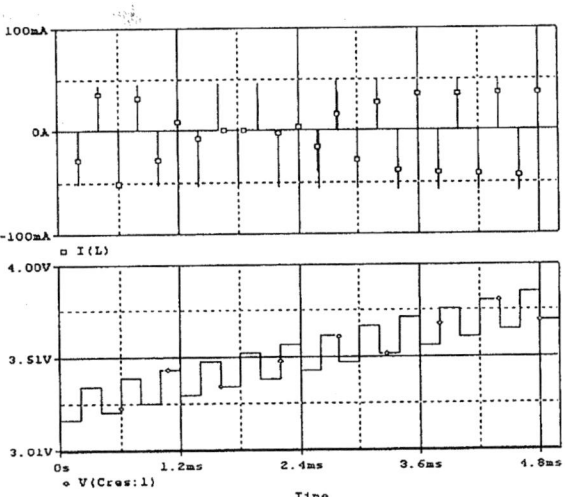

Fig.7: i_L and V_{Cres} over a multiple cycles of the input vibration

Figure 8 shows a timing diagram which represents the locations of the required controlling pulses with respect to the input displacement (X). It indicates that four controlling pulses are required to be generated from the controller. The first two controlling pulses are required to perform the charging operation of the electrostatic MEMS converter at C^*_{max} to V_{min} from C_{res}. The other two controlling pulses are required to perform the discharging operation of the electrostatic MEMS converter at C^*_{min} in C_{res}.

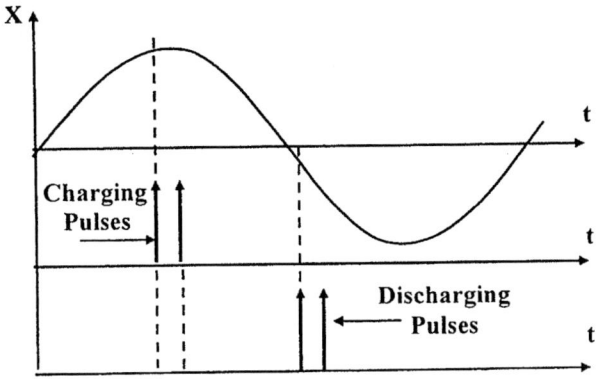

Fig.8: The required four controlling pluses out of the controller with respect to input displacement (X)

V. CONCLUSION

A SIPCE model represents the behavior of the electrostatic MEMS converter is introduced.

C_{res} acts as the supply for the energy scavenging system. It gives the initial charge to the converter at the beginning of the vibration cycle. In addition it gains energy at the end of the cycle from the conversion operation. The required controlling pulses which control the charging and the discharging operations of the converter are determined. Two charging pulses are needed to be generated from the controller to complete the charging of the converter. In addition, two discharging pulses are needed to complete the discharge of the converter.

VI. FUTURE WORK

The design of the controller circuit of the electrostatic MEMS converter, depending on the determination of the required charging and discharging pulses of the converter will be introduced.

REFERENCES

[1] S. Roundy, P. K. Wright, and J. Rabaey, "A study of low level vibrations as a power source for wireless sensor nodes", Computer Communications, vol. 26, 1131-1144, 2002.

[2] J. Rabaey et al., "Pico-Radio Supports Ad Hoc Ultra-Low Power Wireless Networking", IEEE Computer, Vol. 33, No. 7, pp. 42-48, July 2000.

[3] S. Roundy, P. K. Wright and K. S. J. Pister, David A. Dornfeld "Energy Scavenging for Wireless Sensor Nodes with a Focus on Vibration to Electricity Conversion", A dissertation of Philosophy in Engineering-Mechanical Engineering in the UNIVERSITY OF CALIFORNIA, BERKELEY, Spring 2003.

[4] Mona S. Salem & Marwa S. Salem, A. A. Zekry and H. F. Ragai, "Design Methodology and Power Conditioning Circuit of Vibration Based MEMS Converters for Wireless Sensor Networks", ITI 3rd International Conference on Information and Communication Technology (ICICT 2005), 5 – 6 December 2005, Cairo, EGYPT

[5] Scott Meninger, Jose Oscar Mur-Miranda, Rajeevan Amirtharajah, Anantha P. Chandrakasan, and Jeffrey H. Lang,, "Vibration-to-Electric Energy Conversion", IEEE TRANSACTIONS ON VERY LARGE SCALE INTEGRATION (VLSI) SYSTEMS, VOL. 9, NO. 1, 64 – 76, FEBRUARY 2001.

Wafer Level Package Using Polymer Bonding of Thick SU-8 Photoresist

Kyounghwan Na, Ill hwan Kim, Eunsung Lee, Hyeon Cheol Kim, Yong-Hwan Lee, Kukjin Chun
School of Electrical Engineering and Computer Science, Seoul National University, Seoul, Korea
na@mintlab.snu.ac.kr

Abstract— For the application to optic devices, wafer level package including spacer with particular thickness according to optical design could be required. In these cases, the uniformity of spacer thickness is important for bonding strength and optical performance. Packaging process has to be performed at low temperature in order to prevent damage to devices fabricated before packaging. And if photosensitive material is used as spacer layer, thickness of spacer and size and shape of pattern can be easy to control. This paper presents polymer bonding with thick, uniform and patterned spacing layer using SU-8 2100 photoresist for wafer level package. SU-8, negative photoresist, can be coated uniformly by spin coater and is cured at 95℃ and bonded well near the temperature. It can be bonded to silicon well, patterned with high aspect ratio and easy to form thick layer due to its high viscosity. It is also mechanically strong, chemically resistive and thermally stable. But adhesion of SU-8 to glass is poor, and in the case of forming thick layer, SU-8 layer leans from the perpendicular due to imbalance to gravity. To solve leaning problem, the wafer rotating system was introduced. Imbalance to gravity of thick layer was cancelled out through rotating wafer during curing time. And depositing additional layer of gold onto glass could improve adhesion strength of SU-8 to glass. Conclusively, we established the coating condition for forming patterned SU-8 layer with 400 μm of thickness and 3.25% of uniformity through single coating. Also we performed silicon to glass bonding with thick, uniform and patterned spacing layer of SU-8 photoresist for wafer level package.

I. INTRODUCTION

For wafer level package of optic devices, hundreds-micron thick and patterned spacer between two substrates can be required according to restriction of optical design, for example, focal length. And the bonding process must be performed at low temperature because most of lens is made with polymers which can be deteriorated at high temperature. The polymer bonding using photosensitive material can satisfy these requirements well. It can be performed at low temperature below 200℃ and allows narrow patterning. And some kind of photosensitive polymer such as SU-8(Microchem), THB 151N(JSR) etc. can form hundreds-micron thick layer due to their high viscosities. SU-8, negative photoresist, is widely used in microfluidics and MEMS parts for several purposes recently because it is chemically resistive, mechanically strong and thermally stable, it can adhere to silicon very well and it can be patterned up to 17:1 of aspect ratio [1]. SU-8 can be used as mold for electroplating [2] or PDMS [3], micro structure such as cantilever [4], gripper [5], or

micro channel [6] and encapsulating material [7]. Here, we used SU-8 as adhesive of Si to glass bonding for wafer level package. Most application of SU-8 mentioned above except mold for electroplating or PDMS is not required hundreds-micron of thickness of SU-8. And Fine control of thickness and uniformity is not very important in cases of uses as mold. But we need very uniform layer of SU-8 for optical performance as well as bonding characteristics. And inclining to one side and poor adhesion to glass are problems of thick SU-8 layer which must be solved.

II. DESIGN

The reasons why we chose SU-8 among several photoresists are mechanical strength, chemical resistivity and good adhesion strength to silicon as mentioned above. The physical properties of SU-8 are described in Table.1. But there is another important reason in choosing SU-8. SU-8 can form 400um film at higher spin speed than others due to its high viscosity. It means we can get more uniform layer, also.

Table. 1 Properties of SU-8 2100

Adhesion Strength (mPa) Silicon/Glass/Glass & HMDS	38/35/35
Glass Transition Temperature (Tg ℃), tan δ peak	210
Thermal Stability (℃ @ 5% wt. loss)	315
Thermal Conductivity (W/mK)	0.3
Coeff. of Thermal Expansion (CTE ppm)	52
Tensile Strength (Mpa)	60
Elongation at break (εb %)	6.5
Young's Modulus (Gpa)	2.0
Dielectric Constant @ 10MHz	3.2
Water Absorption (% 85/85 RH)	0.65

It is known that the adhesion strength of SU-8 to glass is poor [4]. Thus, the experiment for improvement of adhesion strength of SU-8 film to glass wafer have to be performed with the experiment for uniform coating of thick photoresist. Several materials were deposited onto glass wafer as interlayer between glass and SU-8 to improve adhesion strength. Gold, Nickel, silicon dioxide, polycrystalline silicon, silicon nitride and Chrome were chosen as adhesion material and SU-8 coated on silicon wafer was bonded to silicon wafer for reference. Through comparison of bonding strength most effective material was founded out.

1-4244-0899-7/06/$20.00 ©2006 IEEE.

Generally, thick photoresist lean on one side because hot plate surface isn't accurately perpendicular to the gravity and so isn't surface of glass wafer. In case of thin photoresist this phenomenon doesn't happen because viscosity of photoresist is increased by evaporation of most of solvent in photoresist during spin coating and soft bake time of thin photoresist film is short. But viscosity of thick photoresist is not increased by much after spin coating and free stream region beyond the boundary layer of thick film is much wider than that of thin film. These can make far more flow in thick photoresist film than thin photoresist film, and moreover flow volume is increased in proportion to the soft bake time which is increased according to film thickness. The effect of glass surface not perpendicular to gravity can be canceled out by rotating glass wafer. So wafer rotating system as seen in Fig. 1 was made which can rotate wafer continuously during bake time.

Optimal condition for thickness and uniformity of photoresist film was established with varying spin speed, spin time, temperature and time of flattening. And optimal patterning condition is established by profile measurement of photoresist film varying exposure energy and developing time.

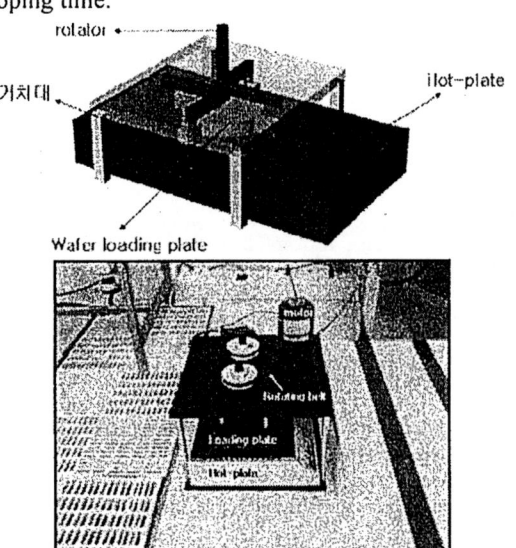

Fig. 1 Wafer rotating system

III. EXPERIMENT

SU-8 2100 of Microchem Co. is used for thick photoresist film. Its properties are described in table. 1. Bare glass wafer was cleaned with acetone/methanol/ isopropyl alcohol for 3 minutes, respectively. Then, the wafer was in $H_2SO_4+H_2O_2(4:1)$ for 10 minutes and was put on hot plate at 200℃ for 10 minutes for dehydration. Dehydration is necessary to improve adhesion between glass wafer and SU-8. And then, coating experiment for 400 μm thickness was performed varying spin speed and spin time. And coating experiment varying temperature and time of flattening was performed for uniformity. Then, spin time and speed were modified again for uniformity. Wafer rotating system was always used during all the curing process. Wafer rotating system consists of motor, rotating belt and rotating plate as seen in Fig. 2. Wafer is put on the rotating plate under which hot plate is and the plate is rotating with the belt which is rotated by motor.

SU-8 3020 was used for adhesion layer test because its component is similar to SU-8 2100 and its coating and patterning condition is well-known. Silicon wafer was patterned for dicing view and glass wafer was deposited by adhesion material. Ti/Au, Ni and Cr were deposited by metal sputtering, silicon oxide and silicon nitride were deposited by PECVD and polycrystalline silicon was deposited by LPCVD. Thickness of each adhesion material was fixed to 3000 Å. SU-8 3020 was coated onto the adhesion layer.

SU-8 layer was baked at 65℃ and 95℃ successively and exposed to light. MA-6 aligner of Karl-suss was used, which has 14mW of power intensity and uses unicolor light with 365nm of wavelength. Exposure time was optimized by inspection of profile of photoresist film. The post exposure bake for crosslinking was performed at 65℃ and 95℃ after exposure. And then the photoresist was developed using SU-8 developer. Every thermal process was heated up and cooling down very slowly.

After that, SU-8 was patterned and the glass wafer was bonded to bare silicon wafer. Bonding condition was decided as 1500kPa of bonding pressure at 100℃ for 10 minutes according to optimization and the other side of bonding was decided as silicon wafer of which adhesion strength is good enough. Samples for measurement of adhesion strength were obtained after dicing the bonded wafers. Fig. 2 shows process flow of this experiment.

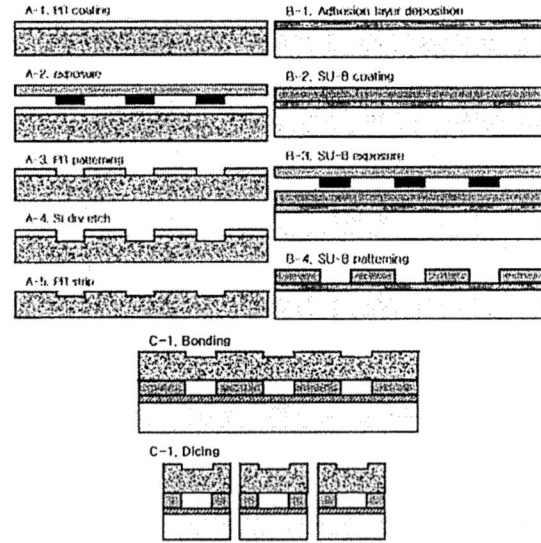

Fig. 2 process flow of experiment

IV. MEASUREMENT AND RESULT

SU-8 film with 400 μm of thickness and 3.25% of uniformity was able to be formed by spin coating with 500rpm 30sec and 780rpm 30sec and then baking for 2 hours at 50℃ and for 2 hours at 95℃ on hot plate through experiments varying spin speed, spin time, flattening temperature and flattening time.

Fig. 3 (a) and (b) shows the distribution of thickness of SU-8 without wafer rotating system and with wafer rotating system, respectively. Graph of Fig. 3 (a) leans on one side, but graph of Fig. 3 (b) is relatively parallel to the surface. These figures say the use of wafer rotating system cancel out the incline due to wafer surface out of perpendicular to the gravity.

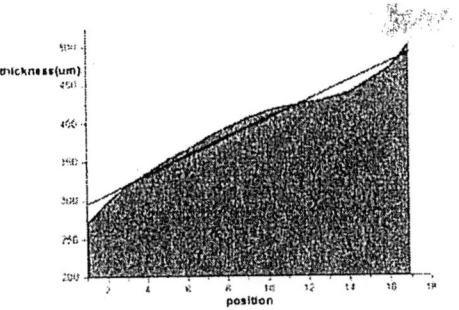

(a) without the wafer rotating system

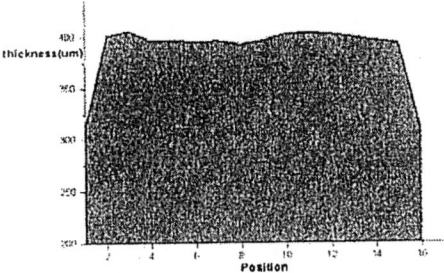

(b) with the wafer rotating system

Fig. 3 Thickness distribution of SU-8 film

As a result of experiment varying exposure time with measurement of SU-8 film profile, 87° of slope was able to be formed by 2400~2800mJ/cm^2 of exposure energy as seen in Fig. 4.

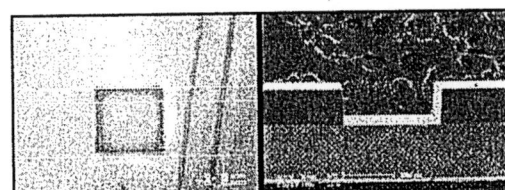

Fig. 4 SEM pictures of patterned SU-8 film

Both pull test for measurement of tensile strength and shear test for measurement of shear strength were performed as measurement of adhesion strength. The method of test is described in Fig. 5.

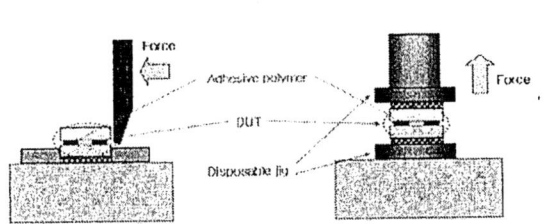

Fig. 5 Schematics of shear test and pull test

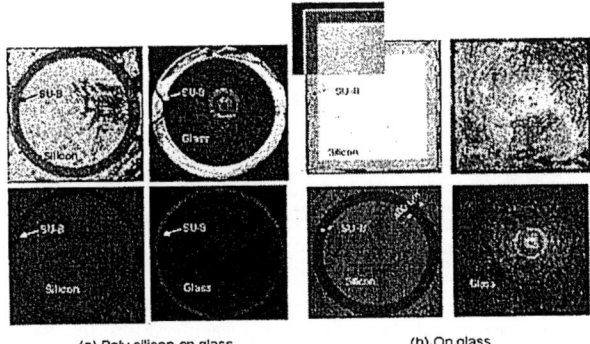

(a) Poly silicon on glass   (b) On glass

Fig. 6 Remainder of SU-8 after tensile test

Pulling chuck with fixed velocity, fracture strength was measured after sample bond to chuck with glue in pull test. In shear test, applying shear force to sample, fracture strength was measured. As seen in Fig.6 (a), when no adhesion layer was used, SU-8 was left only on Silicon side after tensile test. But as using polycrystalline silicon as adhesion layer, SU-8 was left both sides on Silicon and poly-silicon after tensile test, seen in Fig.6 (b). Fig. 7 shows average and maximum value of adhesion strength according to several materials. We can know that the use of gold or polycrystalline silicon as adhesion layer can improve adhesion strength.

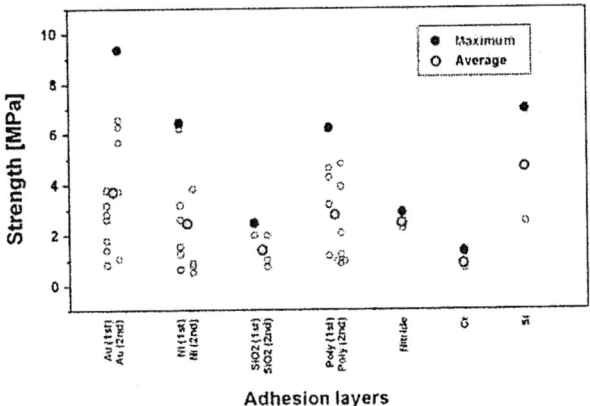

Fig. 7 Tensile strength according to adhesion materials

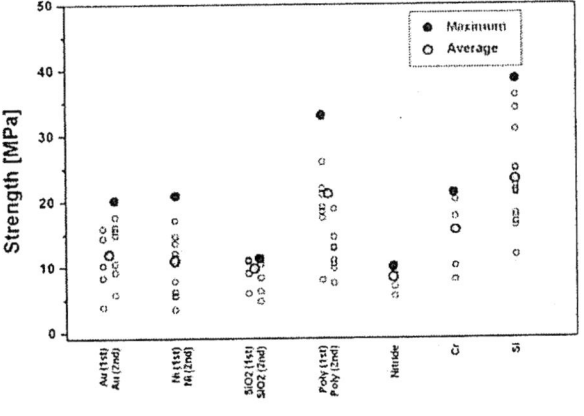

Fig. 7 Shear strength according to adhesion materials

V. CONCLUSION AND DISCUSSION

Using wafer rotating system to eliminate incline of thick SU-8 film, we established the coating condition for forming SU-8 layer with 400 μm of thickness and 3.25% of uniformity through experiments varying spin speed, spin time, flattening temperature and flattening time. And we improved adhesion strength by using gold or polycrystalline silicon as adhesion layer.

We are performing the optimization of wafer bonding using 400 μm-thick SU-8 interlayer and the improvement of adhesion strength based on the results of this paper. In the conference, we'll present the result of bonding test and the optimized condition.

REFERENCES

[1] H. Lorenz, M. Despont, N. Fahrni, J. Brugger, P. Vettiger, P. Renaud, "High-aspect-ratio, ultrathick, negative-tone near-UV photoresist and its applications for MEMS", Sensors and Actuator A, vol. 64, pp. 33-39, 1998

[2] S. Horiguchi, M. Sitti, and H. Hashimoto, "Visualization interface for AFM-based nano-manupulation" ,Proceeding of the IEEE international Symposium on vol. 1, pp. 310-315, 1999

[3] B.-H. Jo, L.M. Van Lerberghe, K.M. Motsegood, D.J Beebe, "Three-dimensional micro-channel fabrication in polydimethyl-siloxane(PDMS) elastomer", Journal of microelectromechanical systems, vol. 9, pp. 76-81, 2000

[4] E. H. Conradie, D. F. Moore, "SU-8 thick photoresist processing as a functional material for MEMS applications", Journal of Micromechanics and Microengineering, vol. 12, pp. 368-374, 2002

[5] N. Chronis and L. P. Lee, "Electrothermally Activated SU-8 Microgripper for Single Cell Manipulation in Solution", Journal of microelectromechanical systems, vol. 14, pp. 857-863, 2005

[6] Y.-J. Chuang, F.-G. Tseng, J.-H Cheng, W.-K Lin, "A novel method of embedded micro-channels by using SU-8 thick-film photoresists", Sensors and Actuator A, vol. 103, pp. 64-69, 2003

[7] P.A. Hammond, D.R.S. Cumming, "Encapsulation of a liquid-sensing microchip using SU-8 photoresist", Microelectronic Engineering, vol. 73-74, pp. 893-897, 2004

Complete Analysis of a Novel Fully Symmetric Decoupled Micromachined Gyroscope

Abdelhameed Sharaf[1,2], Sherif Sedky[2,3], S. E.-D. Habib[4]

[1]NCRRT, EAEA, 3 Ahmed Elzomer Street, Nasr City, Cairo, Egypt
[2]STRC, AUC, 113 Kasr El Eini Street, 11215, Cairo, Egypt
[3]Physics Department, AUC, 113 Kasr El Eini Street, 11215, Cairo, Egypt
[4]Electronics and Communications Dept., Faculty of Engineering, Cairo University, 12613, Giza, Egypt
Email: a_sharaf@aucegypt.edu

Abstract

This paper reports on a novel fully symmetric decoupled micromachined gyroscope (FSDMG). The proposed sensor has five masses, electrostatically driven to primary mode oscillation, and senses, capacitively, the output signal. The new structure achieves complete decoupling between drive and sense modes to minimize the mechanical crosstalk, thanks to the intermediate mass and decoupling beams. The fully symmetric structure helps to lower the zero rate output of the sensor. The manufacturing asymmetry can be overcome by four sets of adjustable electrodes surrounding the outer frame. Drive and sense amplitudes, mechanical and electrical sensitivities, quality factor and approximated bandwidth are extracted analytically and results are confirmed using finite element analysis (FEA). The designed sensor shows drive and sense modes resonance frequencies of 6391 Hz and 6393 Hz respectively; the frequency mismatch is lower than 0.03%. The drive and sense capacitance are 0.098 pF and 0.23 pF respectively. The FEA using ANSYS shows drive and sense frequency of 6187 and 6199 Hz respectively for only 0.2% mismatch in resonance frequency. The output signal achieves 70 nm amplitude for 1 °/s input rotation rate, which results in 0.83 aF change in the sense capacitance. The numerical value of both the mechanical and electrical sensitivities are 0.07 µm/(°/s) and 17.5 mV/(°/s) respectively.

Keywords: Micromachined Gyroscope, MEMS.

1. Introduction

The Gyroscope is an essential component of any navigation system, and it is a sensor used to measure the angular rotation of its host. It has a wide variety of applications ranging from satellite attitude sensing to rollover detection in automobiles [1]. The Gyroscope family is a large one, and it has many types such as mechanical, optical, and vibratory. Gyroscopes can be classified according to their performance to rate grade, tactical grade, or inertial grade [1, 2].

Over the last two decades, MEMS gyroscopes have received lots of attention due to their small size, batch fabrication, Integrated Circuit (IC) compatibility, low cost, and acceptable moderate performance for most applications such as automotive applications [3]. Researchers at HSG-IMIT, Germany, developed a new silicon rate gyroscope called MARS-RR, which means *M*icromachined *A*ngular *R*ate *S*ensor with two *R*otary oscillation modes [4]. The described device yields–random walk and bias stability of 0.14°/h and 65°/h, respectively. Next, they developed a new silicon rate gyroscope, doubly decoupled and LL-structure, based on the principle of DAVED, which means *D*ecoupled *A*ngular *V*elocity *D*etector. This device is characterized by having two linear oscillations, LL-structure, in both excited and sensed modes. In addition, this device decouples completely the drive and sense modes [5]. The MARS-RR has a large area (~ 6mm^2) and complex structure, as the sense electrode must be sandwiched between the drive electrode and the substrate surface. The LL-structure solves these problems by using comb drive sense capacitance scheme to sense the secondary oscillations, but has nonlinear operation in the sense combs and limited dynamic range due to capacitance change in the gap variations. Other types were introduced which discuss the decoupling problem [6], symmetry problem [7], technologies development [8], driving and sensing schemes [9], closed or open loop operations [10], suggest new structures [11], introduce new design principles [12], drive thermal and structural analysis [13], deduce dynamic control analysis [14], and develop new fabrication processes [15].

This work introduces a new design for a triple mass gyroscope, and decoupled mode operation. The proposed structure is based on surface micromachining technology [16] and is simulated using ANSYS software. The nonlinearity of the sense combs introduced above is avoided by using lateral combs, which have the advantages of linear operation, large capacitance, and large dynamic range. The proposed sensor leverages the low thermal budget advantages of pulsed laser deposited silicon germanium [17]. The simulated results show good mechanical and electrical sensitivities of the sensor. They also point out the good resolution of the proposed gyroscope, which is limited by the readout electronics, and the very low aspect ratio of the prposed MEMS technology. Performance can be increased by one order of magnitude, if the sensor is manufactured using high aspect ratio technology.

This paper is organized as follows: Section 2 discusses the gyroscope design, principle of operation, structure design and drives the equations of motion. Section 3 introduces the analytical and finite element results. Finally conclusions are given in section four.

2. Gyroscope Design

2.1 Structure

Figure1 shows a schematic diagram of the proposed gyroscope structure refered to as fully symmetric decoupled micromachined gyroscope (FSDMG). It consists of an outer frame, inner mass, and sense element. The outer frame, which consists of two C-shaped frames, is anchored to the substrate through a crab-leg beam suspension, beams 1, and 2. It carries two combs assembly, and has a mass m_d. The first comb is used as drive actuator, and is assembled as four combs that are arranged symmetrically around the outer frame. The second comb is a control and sense comb. The sense comb is used to sense the drive mode displacement for feedback control purposes. The control combs have a control signal for electrostatic tuning. The inner mass decouples the drive and sense elements and has a mass m_i and it is attached to the outer frame via suspension beam 3, and to the sense element through folded beam, beam 4 (Fig. 1). The sense element has a mass m_s and consists of two masses and supports the sense combs. It is anchored to substrate through suspension beam 5, and to the inner mass via

simple beam (beam 4 in the figure 1). The mechanical crosstalk is minimized due to the decoupling between the drive and sense masses.

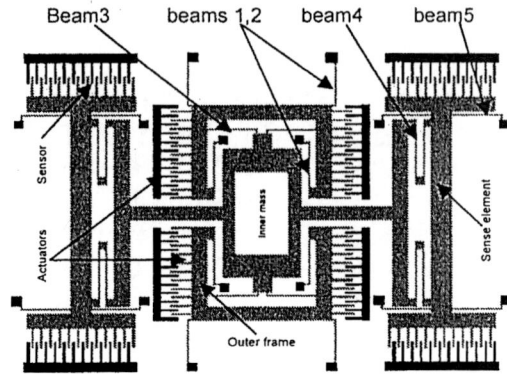

Figure 1 schematic diagram shows the main parts of the FSDMG

2.2 Principle of Operation

The principle of operation of all vibratory MEMS gyroscopes is based on transferring energy between two modes of oscillation. Appling an appropriate ac voltage across the drive comb-finger produce an electrostatic force. This force is a function of the number of comb fingers, structure thickness, applied potential difference and the gap separation between fingers [18]. If the drive force is at the resonance frequency of the drive mode (primary mode) the drive mode displacement is amplified by the mechanical quality factor of the drive mode. When exposed to external rotation rate around the z-axis, an induced Coriolis force is produced along the y-axis. This force is a function of the drive mode velocity, the external rotation rate and the sense mass [19]. Accordingly, maximizing the drive mode amplitude and the structure mass is required to sense small input rotation rates. The Coriolis force excites the sense mode oscillation (secondary mode along the y-axis). If the sense element is at the resonance frequency of the drive mode (matched mode operation) the sense displacement amplitude is amplified by the mechanical quality factor of the sense mode.

2.3 Design Principles

The design process begins by sizing the drive mode to achieve visible oscillations in the drive mode amplitude. The internal and external elements must be sized simultaneously to keep the resonance frequencies of the two modes matched. Driving the sensor to primary mode oscillation is done electrostatically by means of a set of comb-fingers. Sensing the drive mode displacement amplitude is done capacitively by means of a set of comb-fingers. The secondary mode displacement is sensed capacitively by means of two sets of comb fingers connected as a voltage divider.

For the crab-leg suspension, we have spring constants in both directions, but it is more compliant ialong one direction and very stiff along the other direction [18].

$$K_r = \frac{E.h.w_b^3.(l_a + 4.l_b)}{l_b^3.(l_a + l_b)},$$

The comb-drive can be used both producing an electrostatic force or used as a sense capacitor. When it is used as an actuator the produced force from a comb-drive actuator is [18]:

$$F_d = \frac{1.14 \, N.\varepsilon_0.h.V^2}{g_0}$$ which results in a

drive mode displacement of $X_d = \dfrac{Q_x.F_d}{K_x}$

if the drive force is at the resonance frequency of the drive mode. If the sensor is exposed to a constant rotation rate along the z-axis, a Coriolis force is induced along the y-axis, and can be expressed as [18]:

$$F_c = -2.m.\Omega \times \dot{X} = \frac{-2.m.\Omega.\omega_x.Q_x.F_d}{K_x}\sin(\omega_x.t).$$

The mechanical sensitivity of the sensor is given

by [reference]: $S_m = \left|\dfrac{y}{\Omega}\right| = \dfrac{2.m.\omega_x.Q_x.Q_y.F_d}{K_x.K_y}$. The

sense combs are designed to have an overlapping length of 20 μm. Using voltage potential divider as shown in Figure 2, the output signal is computed as:

$$V_o = \frac{C_2}{C_1 + C_2}V_m = \frac{V_m}{2} + \frac{y.V_m}{2.l_o}.$$ This yields an

electrical sensitivity of 0.25 V/μm for a carrier signal amplitude of 10 V.

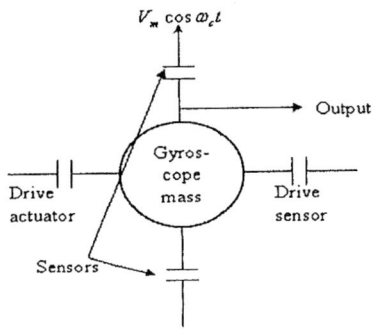

Figure 2: The main gyroscope components and output connection

3. Results

3.1 Analytical Results

The material used is $Si_{0.7}Ge_{0.3}$, which has Young's modulus of 91.06 GP, density of 3332 Kg/m³, and Poisson ratio of 0.2765. Using the above constants with the aid of the stated equation in section 2, the spring constants are 8.3 N/m and 9.99 N/m in the x- and y-directions respectively. The drive actuator can produce an electrostatic force of 0.91μN; this force results in a drive mode displacement of 5.48μm, at a drive mode quality factor of 50. To maximize the sensor sensitivities, it must be operated under matching conditions. The designed masses have values of 1.5827 μg for outer frame, 3.5652 μg for inner mass, and 1.3128 μg for the sense element. The total drive mass is 5.1479 μg, and the sense mass is 6.1908 μg, and the total mass of the structure is 7.7735 μg. These values result in resonance frequencies of 6391 Hz and 6393 Hz for the drive and sense mode respectively. The frequency mismatch is as low as 0.03%. The induced Coriolis force has an amplitude of 2.73 nN/(°/s), resulting in sense mode displacement of 13.6 nm, at sense mode quality factor of 50. The mechanical and electrical sensitivities are 0.014μm/ (°/s) and 3.5 mV/ (°/s) respectively. The output is linear up to ± 320 °/s. It is instructive to refer to table 1 which summarizes all the analytical constants and results.

Table 1 FSDMG analytical results

Material	$Si_{0.7}Ge_{0.3}$
Young's Modulus	91.06 GPa
Density	3332 Kg/m³
Poisson Ratio	0.277
X- direction spring	8.3 N/m
Y-direction spring	9.99 N/m
Drive force	0.91 μN
Drive amplitude	5.48 μm
Outer frame mass	1.5827 μg
Inner mass	3.5652 μg
Sense element mass	1.3128 μg
Drive mass	5.1479 μg
Sense mass	6.1908 μg
Drive frequency	6391 Hz
Sense frequency	6393 Hz
Mismatch	0.03 %
Coriolis force	2.73 nN/(°/s)
Sense amplitude	13.6 nm
Mechanical sens.	0.014 μm/(°/s)
Electrical sens.	3.5 mV/(°/s)
Dynamic range	±320 °/s

3.2 Finite Element Analysis

Figures 3 through 6 show the modal and harmonic analysis for the FSDMG as obtained using ANSYS. Figure 3 shows the drive mode shape with natural frequency of 6187 Hz. It has a translational mode shape displacement as being designed.

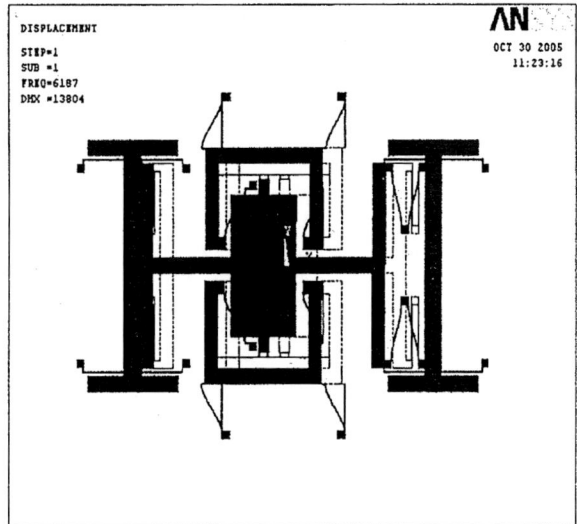

Figure 3: Drive mode shape

Figure 4 shows the secondary mode shape, which has also a translational mode shape in the y-direction at a resonance frequency of 6199 Hz. The frequency mismatch is only 0.2%.

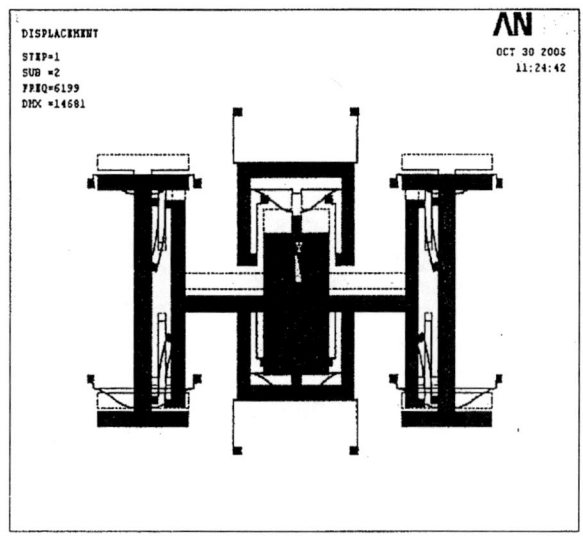

Figure 4: Sense mode shape

Figure 5 shows the frequency response of the drive's outer frame. The vertical axis has two different scales. This mass is designed to have dominantly x-direction displacement. The curve shows a maximum x-displacement of 5.533 μm at resonance, and a small y-displacement of 1.355 nm at resonance, which indicates a weak coupling between the two modes. Figure 6 shows the frequency response of the sense element. The curve shows a maximum x-displacement of 10.4 nm at resonance, and a large y-displacement (69.6 nm) at resonance. Note that this element is forced to move only in the sense mode direction (y-direction). These values are computed at an input rotation rate of 1 ($^{\circ}$/s).

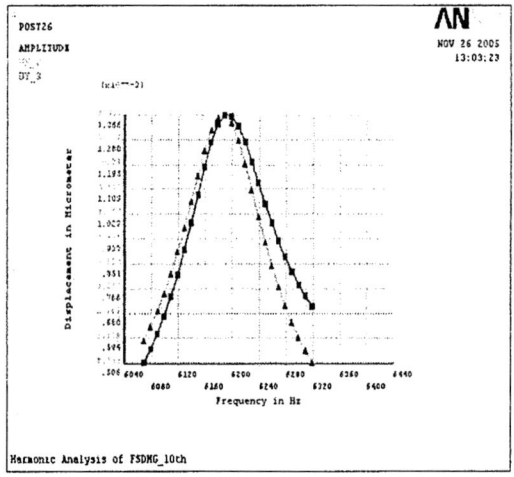

Figure 5: Drive Element frequency responses

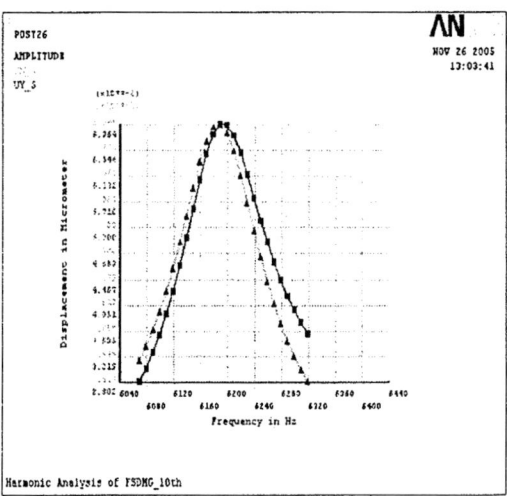

Figure 6: Sense Element frequency response

The drive mode Q-factor is about 51.6, and that of the total sense mode quality factor is about 47.7. The mechanical sensitivity is 0.07 μm/ (°/s). The electrical sensitivity can be computed to be 17.5 mV/ (°/s). Note that the simulation results are in good agreement with the analytical calculation.

It is instructive to refer to Table 2, which summarizes the analytical and numerical analysis for this structure.

Table2 ANSYS modal and harmonic analysis results

	Analytical	Numerical
Area (mm^2)	1.694	1.694
Drive freq. (Hz)	6391	6187
Sense freq. (Hz)	6393	6199
x_d (μm)	3.144	5.533
Y_s (μm)	0.014	0.069
Mechanical sens. (μm/(°/s))	0.014	0.07
Electrical sens. (mV/(°/s))	3.5	17.5
Total mass (μg)	7.7735	7.78
Dynamic range (°/s)	250	150
Bandwidth (Hz)	~	~ 120
Quality factor, Q_x	Assumed 50	~ 51.6
Quality factor, Q_y	Assumed 50	~ 47.7

4. Conclusions

This paper proposed a novel structure for decouple vibratory MEMS gyroscope. It is a triple mass type, electrostatically driven to resonance and capacitively senses the output. Decoupling both modes of oscillations minimizes mechanical cross talk. The fully symmetric design helps to minimize the zero rate output (ZRO). We are currently investigating the possibility of increasing the sensor perfromace.

References

1. Navid Yazdi, et.al., "Micromachined Inertial Sensors," Proc. Of the IEEE, Vol. 86, No. 8, August 1998, PP 1640-1659.
2. A. Lawrence, "Modern Inertial Technology," Springer Verlag, 1993.
3. Huikai Xie, Gary K. Fedder, "Integrated Microelectromechanical Gyroscopes," J. of Aerospace Engineering ASCE/April 2003, PP 65-75.
4. W. Geiger, et.al., "A new silicon rate gyroscope," Sensors and Actuators A 73, 1999, PP. 45-51.
5. W. Geiger, et.al., "Decoupled microgyros and the design principle DAVED," Sensors and actuators A 95, 2002, PP. 239-249.
6. Said E. Alper, Tayfun Akin, "A single-Crystal silicon symmetrical and decoupled MEMS gyroscope on an insulating substrate," J. of Microelectromechanical systems, Vol. 14, No. 4, August 2005, PP. 707-717.
7. Said E. Alper, Tayfun Akin, "A symmetric surface micromachined gyroscope with decoupled oscillations," Sensors and Actuators A 97-98, 2002, PP. 347-358.
8. Farrokh Ayazi, et.al., "A HARPSS poltsilicon vibrating ring gyroscope," J. of Microelectromechanical systems, Vol. 10, No. 2, June 2001, PP. 169-179.
9. Heng Yang, et.al., "Two-dimensional excitation operation mode and phase detection scheme for vibrating gyroscope," J. Micromech. Microengi. , Vol. 12, 2002, PP. 193-197.
10. S. An. Et.al., "Dual-axis microgyroscope with closed-loop detection," Sensors and Actuators A 73, 1999, PP. 1-6.
11. A. Sharaf, Sherif Sedky, Serag E.-D. Habib, "Design and Simulation of a New Decoupled Micromachined Gyroscope," Journal of Physics: Conference Series 34 (2006) pp. 464 469.
12. Sangwoo Lee, et.al., "Surface/Bulk Micromachined Single-Crystalline-Silicon Micro-Gyroscope," J. of Microelectromechanical systems, Vol. 9, No. 4, Dec. 2000, PP. 557-567.
13. C. C. Painter et.al., "Structural and Thermal modeling of a z-axis rate integrating gyroscope," J. Micromech. Microengi., Vol. 13, 2003, PP. 229-237.
14. Andrei M. Shkel, et.al., "Dynamics and control of micromachined gyroscopes,"

15. Proc. Of the American Control Conference. San Diego, California, June 1999, PP. 2119-2124.

16. Srinivas A. Tadigadapa, Nader Najafi, "Developments in Microelectromechanical systems (MEMS): A manufacturing Perspective," Transactions of the ASME, Vol. 125, Nov. 2003, PP. 816-823.

17. http://www.memscap.com/memsrus/cr mumps.html.

18. Sherif Sedky, Ibrahim El Deftar, and Omar Mortagy, "Pulsed Laser Deposition of Boron Doped $Si_{70}Ge_{30}$", Materials Research Society Meeting, Spring 2006, Vol. 901, A12-02, April 16-20, 2006.

19. William C. Tang, et.al., "Electrostatic Comb-drive of Lateral Polysilicon Resonators," Sensors and Actuators A 21-23, 1990, PP. 328-331.

20. Mickael Kranz, "Design Simulation and Implementation of Two Novel Micromachined Vibratory-Rate Gyroscope," MSc. Thesis, Carnegie Mellon University, May 1998.

21. William C. Tang, et.al., "Laterally Driven Polysilicon Resonant Microstructures," Sensors and Actuators A 20, 1989, PP. 25-32

A New Design of a Current-mode Wheatstone Bridge Using Operational Floating Current Conveyor

Yehya H. Ghallab

Biomedical Eng. Dept., Helwan University, Cairo, Egypt
Department of Electrical and Computer Engineering
University of Calgary, 2500 university drive, N.W,
Alberta, T2N 1N4, Canada..
yghallab, @.ucalgary.ca

Wael Badawy

Department of Electrical and Computer Engineering
University of Calgary, 2500 university drive, N.W,
Alberta, T2N 1N4, Canada.

badawy@.ucalgary.ca

Abstract— **This paper presents a new topology for current-mode Wheatstone bridge (CMWB) that utilizes an Operational Floating Current Conveyor (OFCC) as a basic building block. The proposed CMWB has been analyzed, simulated, implemented and experimentally tested. The experimental results verify that the proposed CMWB out performs existing CMWBs in terms of accuracy.**

I. INTRODUCTION

Traditional voltage-mode Wheatstone bridge (VMWB) offers a good method for measuring small resistance changes accurately. Therefore, the Wheatstone bridges are used for sensing temperature, strain, pressure, fluid flow, and dew point humidity [1, 2].

Recently, a method based on circuit duality concept [3] is developed and achieved a current-mode Wheatstone bridge (CMWB) [4]. The advantage of the proposed CMWB are: (1) reduction of sensing passive elements, i.e. we can use two resistors instead of four, and getting the same performance, (2) superposition principle can be applied without adding any signal conditioning circuitry, thus addition of sensor effects is possible, and (3) it has a higher common-mode cancellation.

In [4], two different topologies to implement a CMWB have been used; the first one used two second-generation current conveyors (CCII), one of them is positive CCII+, and the second is negative CCII- (see Fig.1). Fig.1 shows the equivalent circuit taking into consideration the equivalent input resistance at X terminal (Rx) of the CCIIs. Assuming that and, and by using a routine circuit analysis, we can prove that the output current Io is related to the reference current Iref by equation (1):

$$I_{out} = I_1 - I_2 = \frac{\pm \Delta R}{R_o + R_x} . I_{ref} \qquad (1)$$

From (1), we can observe that I_{out} is inversely proportional to Rx, and the accuracy is limited by the tolerance of Rx, which is low. Also, the linearization topology used, when only one resistor is varying, i.e. and is shown in Fig.1. In this case the output current Io is related to the reference current Iref by:

$$I_o = I_x = \frac{\pm \Delta R + R_x}{R_o + R_x} . I_{ref} \qquad (2)$$

Therefore, the disadvantages of using CCII as a building block in a CMWB are the limited accuracy [see (1), and (2)]

and the need of more circuitry for linearization.

In this paper a new CMWB topology has been proposed. It has the combined advantages of all these CMWBs. In other words, it is simple and reduces the sensing passive elements, as we have used only two resistors instead of four and getting the same performance. Also, we can use the principle of superposition without adding any signal conditioning circuitry. It has much improved common-mode cancellation and furthermore it has high accuracy.

The proposed CMWB topology is based on the OFCC, which exhibits flexible characteristics with respect to other current-mode or voltage-mode devices [5, 6]. The remainder of this paper is organized as follows:

Section II introduces and reviews the basic concept of the OFCC and its characteristics. A detailed analysis of the proposed CMWB is presented in Section III. Section IV represents the simulation and experimental results. Section V concludes the paper and discusses the merits of the proposed CMIA based on the experimental findings.

II. THE OPERATIONAL FLOATING CURRENT CONVEYOR (OFCC)

The OFCC is a five-port network, comprised of two inputs and three output ports, as shown in Fig.2. In this diagram, the port labeled X represents a low impedance current input, port Y is a high impedance input voltage, W is a low impedance output voltage, and Z+, and Z- are the high impedance current outputs with opposite polarities. The OFCC operates where the input current at port X is multiplied by the open loop transimpedance gain Zt to produce an output voltage at port W. The input voltage at port Y appears at port X and thus a voltage tracking property exists at the input port. Output current flowing at port W is conveyed in phase to port Z+ and out of phase with that flowing into port Z-, so in this case a current tracking action exists at the output port. Thus, the transmission properties of the ideal OFCC can be conveniently described as:

$$\begin{bmatrix} i_y \\ v_x \\ v_w \\ i_{z+} \\ i_{z-} \end{bmatrix} = \begin{bmatrix} 0 & 0 & 0 & 0 & 0 \\ 1 & 0 & 0 & 0 & 0 \\ 0 & Z_t & 0 & 0 & 0 \\ 0 & 0 & 1 & 0 & 0 \\ 0 & 0 & -1 & 0 & 0 \end{bmatrix} \begin{bmatrix} v_y \\ i_x \\ i_w \\ v_{z+} \\ v_{z-} \end{bmatrix} \quad (3)$$

where i_y and v_y are the inward current and voltage at the Y port, respectively, as shown in Fig.2. i_x and v_x are the input current and voltage at the X port, respectively. i_w and v_w are the output current and voltage at W port, respectively. i_{z+} and v_{z+} are the output current and voltage at Z+ port, respectively. Similarly, i_{z-} and v_{z-} are the output current and voltage at the Z- port, respectively. Z_t represents the impedance between X and W ports.

The OFCC can be implemented by applying the principle of supply current sensing to a current feedback (CFB) op-amp [7] such as illustrated in Fig.3. The current mirrors CM1 and CM2 establish the output current at port Z+. Also, CM1 and CM2 and their cross-coupling with the current mirrors CM3 and CM4 through the current steering transistors CS1 and CS2 generate a complementary output current at port Z-. The OFCC is basically designed to be used in a closed loop configuration, with current being fed back from port W to port X.

III. THE PROPOSED CMWB

The proposed CMWB consists of three operational floating current conveyors (OFCC), three feedback resistors (R_{W1}, R_{W2} and R_{W3}), two sensitive resistors (R_1 and R_2) (one or both of them can represent a sensor), and a ground load (R_L), as shown in Fig.4. The passive components used in OFCC design are given in Table 1. It may be observed that both R_{z+} and R_{z-} are very large and therefore may be neglected especially when these resistors act in parallel with a significant smaller load (R_L). Furthermore, the input resistance at X terminal (R_X) is very small (<0.002Ω) and hence can be neglected (see [8], [9] for detail analysis of the calculation of R_X)

Taking into consideration both the voltage and current tracking errors of the OFCC, the current tracking error between ports X, Z+ and Z- is:

$$\alpha = 1 - \varepsilon_+ \quad (4)$$

and

$$\gamma = 1 - \varepsilon_- \quad (5)$$

where: ε_+ and ε_- denotes the finite current tracking error at the high impedance output Z+ and Z-, respectively. Thus port currents may then be expressed as $I_{z+} = \alpha I_x$ and $I_{z-} = \gamma I_x$. As we mentioned before that there is a voltage tracking between X and Y terminals of the OFCC, so $V_A = V_B = 0$. The current I_1 and I_2 are calculated as follows:

$$I_1 = \frac{R_2}{R_1 + R_2} \cdot I_{in} \quad (6)$$

$$I_2 = \frac{R_1}{R_1 + R_2} \cdot I_{in} \quad (7)$$

The input current I_{in} is:

$$I_{in} = I_1 + I_2 \quad (8)$$

The output terminals currents from the OFCC (1), I_3 and I_4 can be calculated as follows:

$$I_3 = \alpha_1 I_{in} \quad (9)$$

$$I_4 = \gamma_1 I_{in} \quad (10)$$

The output current from the OFCC (2), I_5 is:

$$I_5 = \gamma_2 I_1 \quad (11)$$

From (9), and (11), I_6 can be obtained

$$I_6 = I_3 + I_5 = \alpha_1 I_{in} + \gamma_2 I_1 \quad (12)$$

The output current from the OFCC (3), I_7 is computed:

$$I_7 = \alpha_3 I_2 \quad (13)$$

The summation of I_3 and I_7 which is I_8 can be obtained from (10) and (13):

$$I_8 = I_4 + I_7 = \gamma_1 I_{in} + \alpha_3 I_2 \quad (14)$$

The output current (I_{out}) is the difference between I_8 and I_6, therefore, we can calculate I_{out} from (12) and (14):

$$I_{out} = I_6 - I_8 \quad (15)$$

$$I_{out} = \alpha_1 I_{in} + \gamma_2 I_1 - \gamma_1 I_{in} - \alpha_3 I_2 \quad (16)$$

For ideal OFCCs, $\alpha_1 = \alpha_2 = \alpha_3 = \gamma_1 = \gamma_2 = \gamma_3 = 1$, thus the output current is:

$$I_{out} = I_1 - I_2 \quad (17)$$

From (6) and (7) into (17), we can get I_{out} in terms of R_1 and R_2:

$$I_{out} = \frac{R_2 - R_1}{R_1 + R_2} \cdot I_{in} \quad (18)$$

Thus, if we have $R_1 = R_0 \mp \Delta R$ and $R_2 = R_0 \pm \Delta R$, then I_{out}

$$I_{out} = \frac{\pm \Delta R}{R_0} \cdot I_{in} \quad (19)$$

So, from (19), we can observe the linear relationship between I_{out} and ΔR, and it doesn't depend on Rx any more [see (8)]. Therefore the accuracy is improved when compared with the CMWB based CCII in [4]. Also, from (19), we can observe that the proposed topology has a high common-mode rejection (CMR) for the common mode current, i.e. $I_1 = I_2$. In other words, when R_1 is equal to R_2, then I_1 is equal I_2 (common-mode current) and the output current $I_{out} = 0$. Also, as we have an output current, hence, the superposition principle can be applied. Thus, it's possible to add the effects of any number of sensors without adding any other circuitry, and this is the great advantage over the voltage-mode Wheatstone bridge VMWB. Moreover, we are using only two sensitive resistors to get the same performance, instead of four as in traditional VMWB.

IV. EXPERIMENTAL AND SIMULATION RESULTS OF THE PROPOSED CMWB

To verify the operational characteristics of the proposed CMWB, the circuit of Fig.4 was simulated using PSPICE version 7.1. The proposed CMWB was also prototyped and the simulation results verified. Each OFCC was constructed using an Analog Devices *AD846AQ* current feedback op amp [10] and current-mirrors composed of Harris transistor array *CA3096CE* [11]. The *AD846AQ* has a bandwidth of 80 MHz at unity gain, and slew rate 450V/µs.

A. The differential measurements

To measure the differential characteristics of the proposed CMWB, we connected the input voltage to V_{in}. Resistors R_{w1}, R_{w2}, R_{w3}, R_L, and R_1 were set at $1K\Omega$ and R_2 was tested at different values ($1.5K\Omega$, $2K\Omega$, $3K\Omega$, and $4K\Omega$). All resistors have 1% tolerance. Fig.13 shows both the experimental as well as the simulation results of the output current (I_{out}) against the input voltage (V_{in}) for R_1 kept constant at $1k\ \Omega$ and R_2 varied. From Fig.5, we can observe that the experimental results validate the simulated results and the analytical results of (19). The difference between the experimental and simulation results can be interpreted as a result of tracking errors and the tolerance of the resistors.

The AC performance of the proposed CMWB, is also tested by connecting an AC source at v_{in} (v_{in} =200mV) at different values of R_2. Fig.6 shows both the experimental as well as the simulation results of the i_{out} against frequency for R_1 kept constant at $1k\ \Omega$ and R_2 is varied. We can observe from this figure that the experimental results are in good agreement with the simulation results, except at frequencies approaching the bandwidth of the OFCC. The difference between the experimental and simulation results can be interpreted as a result of tracking errors and the presence of additional stray capacitances at the various nodes in the circuit. Also, we can observe that the bandwidth (50Meg Hz) is high and constant with different R_2 values.

B Common-mode measurements

To measure the common-mode rejection (CMR) of the circuit in Fig.7, we have selected $R_1=R_2=1K\Omega$. CMR was measured experimentally as a function of frequency. The result obtained is plotted in Fig. 7. From this figure, we can see that the proposed topology exhibits a very small common-mode current, which is independent on the frequency. This current can be interpreted as a result of tracking errors.

V. CONCLUSION

A new CMWB topology based on an OFCC has been proposed, simulated and prototyped. The experimental results show that the new CMWB configuration has the following advantages: The proposed CMWB is not complicated, as it is based on a current-mode device (i.e. OFCC), which shows flexible properties with respect to other current or voltage-mode circuits. Also, we can add the sensor effects, superposition ability, without using complicated circuitry and this is the excellent advantage over the traditional VMWBs. Furthermore, we can reduce the number of sensing passive elements, i.e. we can use two resistors instead of four, and getting the same performance as the VMWBs. So, the cost of using more signal processing equipment will be decreased. The output current, of our CMWB is independent on R_x of the current feedback op-amp used and dependent only on the

external resistors. Moreover, it would be suitable candidate for integration in an IC process. Thus, it can be used in many applications, such as biomedical and lab-on-a-chip.

VI. ACKNOWLEDGEMENT

The authors want to acknowledge National Science and Engineering research Council (NSERC) strategic grant, STPGP 258024-02, Canadian Microelectronics Corporation (CMC), Micralyne for funding this work.

REFERENCES

[1] Reinaldo J. Perez, Design of Medical Electronic devices, Academic press, USA, 2002.

[2] Joseph J. Carr, and John M. Brown, Introduction to Biomedical Equipment Technology, John Wiley and Sons, USA, 1981.

[3] L. P. Huelsman, Basic Circuit Theory, 3rd Edition, Prentice Hall, USA, 1991.

[4] S. Azhari, and H. Kaabi, " AZKA Cell, the Current-Mode Alternative of Wheatstone Bridge," IEEE Trans. Circuits and Systems-I, Vol. 47, No. 9, pp. 1277-1284, 2000.

[5] Y. H. Ghallab, M. Abo El-Ela and M.Elsaid, "Operational Floating Current Conveyor: Characteristics, Modeling and Experimental results," ICM99, Kuwait, 1999.

[6] Yehya H. Ghallab, Wael Badawy, M. Abou El-Ela, and M. H. El-Said" The Operational Floating Current Conveyor and Its Applications", Journal of Circuits, Systems and Computers, Vol. 15, No. 3, pp. 352-371, June 2006.

[7] S. Soclof, "Design and Applications of Analog Integrated Circuits", Englewood Cliffs, N. J Prentice Hall Inc. Chap.9, pp.443-460, 1991.

[8] Yehya H. Ghallab, Wael Badawy, Karan V.I.S. Kaler and Brent J. Maundy," A Novel Current-Mode Instrumentation Amplifier Based on Operational Floating Current Conveyor", IEEE Transaction on Instrumentation and Measurement, vol. 54, no.5, pp. 1941-1994, October 2005.

[9] Yehya H. Ghallab, and Wael Badawy " A New Topology for a Current-mode Wheatstone Bridge" IEEE Transaction on Circuit and System II, vol. 53, no.1, pp. 18-22, January 2006.

[10] Analog Devices Manual "450 V/µs, precision, current-feedback OpAmp (AD846)" pp. (2-307)-(2-317).

[11] Harris semiconductor "CA3096, CA3096A, CA3096C, NPN transistor arrays" File Number 595.4, December 1997.

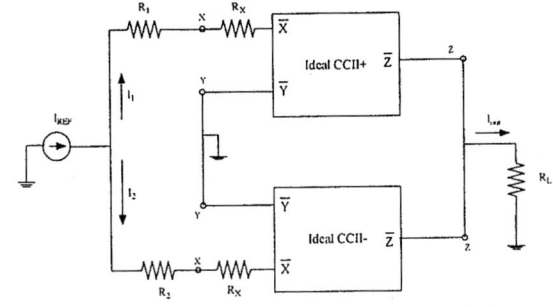

Fig.1 Practical CMWB based on the equivalent circuit of CCII

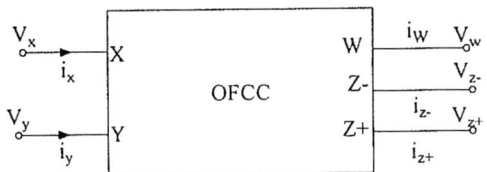

Fig.2 Block diagram representation of the Operational Floating Current Conveyor

The 2006 International Conference on MEMS, NANO and Smart Systems

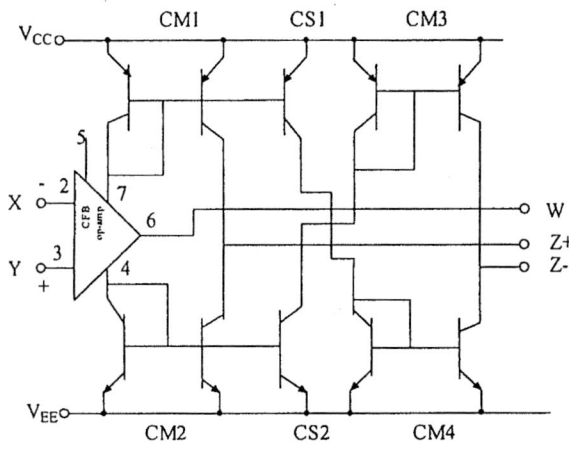

Fig.3 Circuit scheme of the OFCC

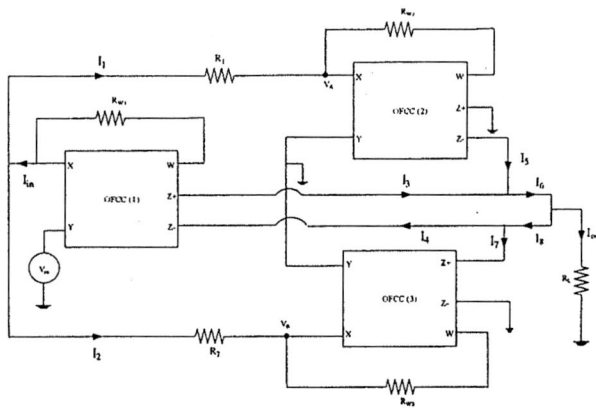

Fig.4 The proposed CMWB based on OFCC

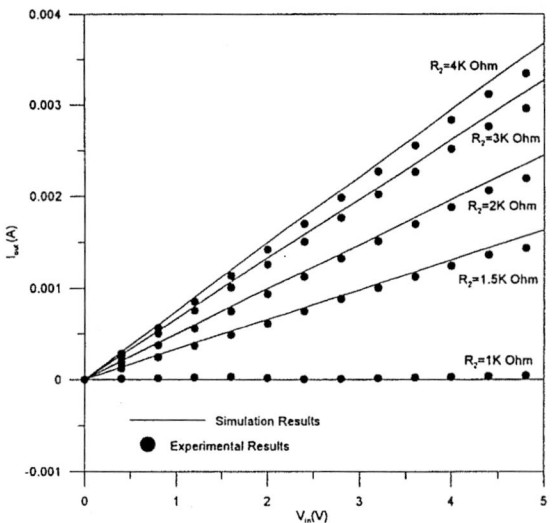

Fig. 5 The Dc response of the proposed CMWB with $R_1=1K\ \Omega$ and R_2 varies

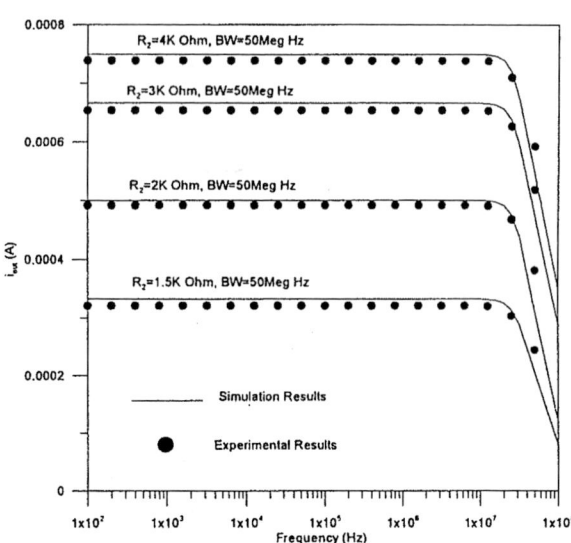

Fig.6 The Frequency response for the proposed CMWB with $R_1=1K$ Ohm and R_2 varies

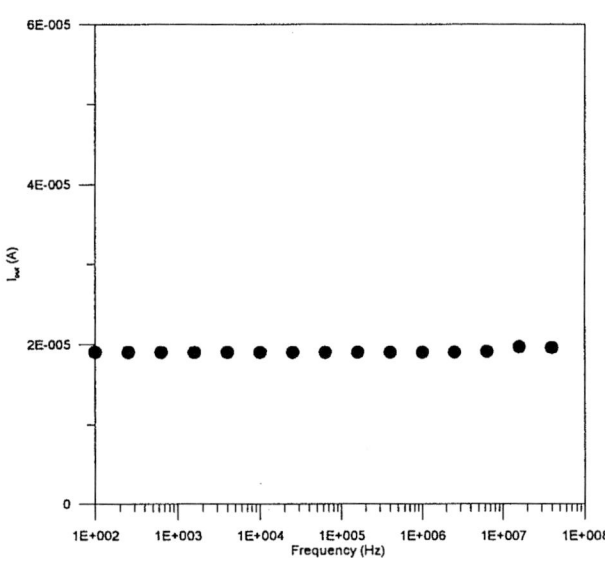

Fig. 7 CMRR for R1=R2=1K Ω

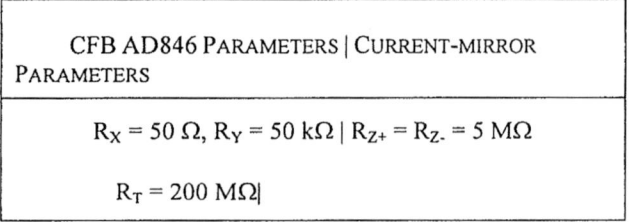

CFB AD846 PARAMETERS	CURRENT-MIRROR PARAMETERS		
$R_X = 50\ \Omega$, $R_Y = 50\ k\Omega$	$R_{Z+} = R_{Z-} = 5\ M\Omega$		
$R_T = 200\ M\Omega$|			

Table 1 OFCC DC parameters (*AD846AQ and CA3096CE*)

44

Performance Enhancement of Gap Closing Electrostatic MEMS Converters

Mona S. Salem*, Marwa S. Salem*, A. A. Zekry*, H. F. Ragai*
*Ain Shams Univ., Fac. of Eng., ECE Dept
mona_marwa2002@yahoo.com

Abstract - This paper proposes new ideas to enhance the output power of the gap closing electrostatic MEMS converters. This will be done by including the effect of the parallel capacitance of the converter in the output power equation. In addition the output power of the converter will be increased by increasing the converter thickness.

List of symbols

Symbol	Definition
E	Output energy from the converter
V_{max}	Maximum allowable voltage for the system
V_{min}	Initial voltage on the converter
C^*_{max}	Total maximum capacitance of the converter
C^*_{min}	Total minimum capacitance of the converter
C_{par}	A constant capacitor, due to anchors, added in parallel to the converter capacitance
f	Input vibration frequency
P_{out}	Output power of the converter
W_a	Anchor width
L_f	Finger length
W_f	Finger width
t	Device wafer thickness
ϵ_0	Permittivity of air
$d_{nominal}$	Nominal distance between fingers
N_g	Number of fixed or movable fingers
L_m	Shuttle mass length
L_t	Total length of the converter
d_{min}	Minimum dielectric gap distance between fingers
A_{anchor}	Anchor area
t_{ox}	Oxide (SiO$_2$) thickness
ϵ_{ox}	Relative permittivity of SiO$_2$
m	Mass of the movable part
ρ	Density of poly Si
W_m	Shuttle mass width
k	Spring constant
E	Modulus of elasticity for poly Si
L_b	Spring shin length
L_a	Spring thigh length
W_{sp}	Spring beam width

I. INTRODUCTION

During the past several years, there has been an increasing interest in the research community on small wireless electronic devices. Wireless sensor networks have many considerable applications in areas ranging from building monitoring and environment control to military applications. Advances in low power Very Large Scale Integration (VLSI) design [1, 2] along with the low duty cycles of wireless sensors have reduced power requirements to the range of tens to hundreds of microwatts. Such low power dissipation opens up the possibility of powering the sensor nodes by scavenging ambient energy from the environment, eliminating the need for batteries and extending the lifetime indefinitely.

Energy scavenging system contains two main parts. The First part is an electrostatic MEMS converter, which converts the environmental vibration into electricity. The other part is the controller circuits, which control the operation of the electrostatic MEMS converter.

There are two well known topologies of electrostatic MEMS converters, gap closing and overlap. The gap closing converter is better than the overlap converter. It gives a higher capacitance range thus higher output power. Therefore, the gap closing converter is more suitable for the energy scavenging applications [1].

This paper presents the output power equation of the gap closing converter. In addition it discusses the different factors which lead to enhancement in the converter output power. These factors include the parallel capacitance of the converter and the thickness of the moving plate.

II. THE OUTPUT POWER EQUATION OF GAP CLOSING CONVERTER

The output energy per cycle from the converter is given by

$$E = \frac{1}{2} V_{max} V_{min} (C^*_{max} - C^*_{min}) \qquad (1)$$

V_{max} is one of the basic constraints for the system which is set by some process and system requirement [3] to 8V and $C^*_{max} = (C_{max} + C_{par})$, $C^*_{min} = (C_{min} + C_{par})$.

1-4244-0899-7/06/$20.00 ©2006 IEEE.

On the other hand to satisfy the charge constrained conversion [3] we have

$$Q = C^*_{max} V_{min} = C^*_{min} V_{max} \qquad (2)$$

The output power from the converter is expressed by

$$P = 2fE \qquad (3)$$

f equals to 2.5 KHz in this case study. The factor of 2 comes from the fact that the converter undergoes two charging/discharging cycles for each mechanical cycle. The power depends on geometric design parameters, physical constraints, and the input voltage. Combining equations (1), (2) and (3), one can derive the following analytical expression for the output power

$$P_{out} = C^*_{min} V^2_{max} (1 - \alpha) f \qquad (4)$$

Where $\alpha = C^*_{min} / C^*_{max}$

Equation (4) represents the output power from the gap closing converter in terms of its capacitances.

III. EFFECT OF THE PARALLEL CAPACITANCE (C_{par}) OF THE CONVERTER ON THE OUTPUT POWER

C_{par} is an important parameter which controls the output power of the converter, as it increases the output energy of the converter thus it increases its output power [3]. Figure 1 shows the MEMS converter cross-sectional view representing C_{par}. The parallel capacitance (C_{par}) is included in the design essentially for free by exploiting the parasitics that exist between the MEMS device and its substrate, and by tailoring the bonding oxide thickness between the two wafers making up the MEMS device.

Since C_{par} acts as a parallel plate capacitor, one has to increase W_a in order to increase C_{par}, thus increasing the output power.

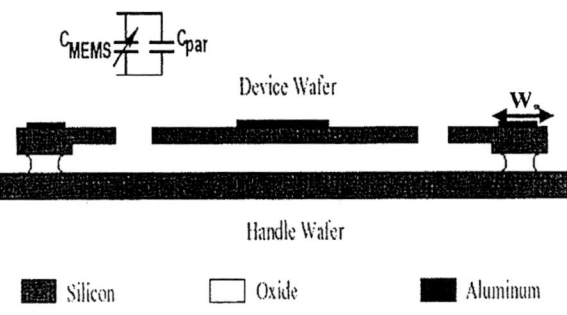

Fig.1: MEMS converter cross-sectional view representing C_{par}

IV. GAP CLOSING CONVERTER DESIGN

In order to calculate the output power of the converter, one has to design the converter dimensions. Figure 2 indicates the designed dimensions of the gap closing converter. It shows that the converter is divided into two parts. The first part is the comb drive. The other part is the spring.

A. Comb Drive Design

The overall chip area of the converter is taken to be 8mm X 8mm, in order to increase the output power of the converter. The effective electrical parameters in the power equation, equation (eq. 4) , C_{max}, C_{min} and C_{par}, are calculated depending on the designed parameters of the converter.

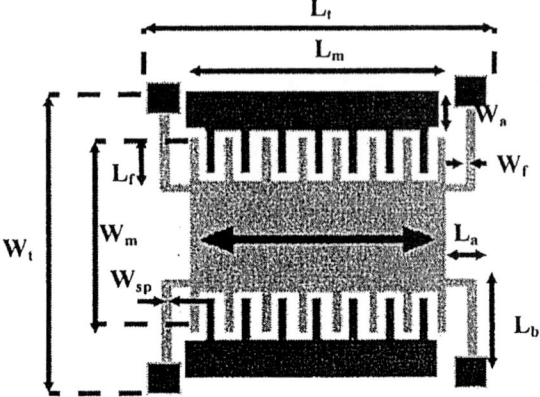

Fig.2: Designed dimensions of gap closing converter

The main objective of the design is to increase the output power of the converter (P_{out}). The converter capacitances are given by the following equations

$$C_{max} = (2N_g \epsilon_0 L_f t) / d_{min} \qquad (5)$$
$$C_{min} = (2N_g \epsilon_0 L_f t) / d_{nominal} \qquad (6)$$
$$C_{par} = (\epsilon_0 \epsilon_{ox} A_{anchor}) / t_{ox} \qquad (7)$$

The designed dimensions of the converter are constrained by the SOIMUMPs technology file [4], which is the chosen technology for implementing the converter.

L_f must be maximized in order to increase the converter capacitance. The maximum value for L_f is limited to 100μm as the converter is anchored at one end only. In addition W_f will be 6μm. t equals to 25μm for SOI technology. ϵ_0 is 8.85 x10^{-12} F/m. It is better to decrease $d_{nominal}$. Therefore N_g is increased, thus increasing the capacitances. So $d_{nominal}$ is set to be 6μm. N_g given by the following equation

$$N_g = L_m / (W_f + 2 d_{nominal}) \qquad (8)$$

L_m is taken to be 7.5mm. The rest of L_t is left for the spring design. Thus N_g is 417 fingers. For d_{min}, it is better to decrease it as much as possible in order to increase C_{max}. Thus d_{min} is taken to be 0.25μm. Substituting in equations (5) and (6), one gets C_{max} equals to 74pF and C_{min} equals to 3pF. Recalling equation (7) to calculate C_{par}, A_{anchor} equals to $W_a L_m$. In addition t_{ox} equals to 1μm from the SOIMUMPs technology file. ϵ_{ox} is 4. Assuming W_a, to be 1.5mm in order to increase C_{par}. C_{par} total for the converter found to be 796.5pF.

Neglecting C_{min} with respect to C_{par}, the power equation (4) becomes

$$P_{out} = C_{par} V^2_{max} (1 - \alpha) f \qquad (9)$$

Since α is given by $C_{par}/ (C_{max} + C_{par})$, then substituting by C_{max}, C_{min} and C_{par} in equation (9), P_{out} is found to be 12.7μW.

B. Spring Design

There are many types of springs. The most suitable one for the gap closing converter design is the crab – leg spring [5]. Figure 3 shows the crab – leg spring with its dimensions.

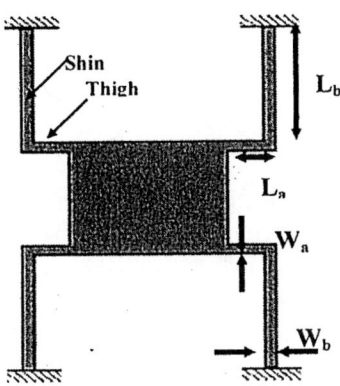

Fig.3: Crab – leg spring

As for the design of the spring, the spring constant must be determined from the following equation

$$f = 1/2\pi \sqrt{(k/m)} \qquad (10)$$

m is given by

$$m = \rho\, t\, W_m\, L_m \qquad (11)$$

ρ equals to 2.33g/cm^3 [6]. W_m is taken to be 4.78mm. Combining equations (10) and (11), k is 515.19N/m. Recalling the crab – leg spring equations [5], the spring constant in X – direction is given by

$$k_x = (EtW_b^3 (4L_b + C\, L_a))/ L_b^3(L_b + C\, L_a) \qquad (12)$$

E is 150Gpa [6]. C is defined as $(W_b / W_a)^3$. Assuming that $W_a = W_b = W_{sp}$, C equals to 1. The spring constant in Y – direction is given by

$$k_y = (EtW_a^3 (L_b + 4C\, L_a))/ L_a^3(L_b + C\, L_a) \qquad (13)$$

From equations (12) and (13), there are three unknowns which are L_b, L_a and W_{sp} thus one must make some assumptions. L_b is assumed to be 1.66mm to be physically feasible. k_y must be larger than k_x to prevent the motion of the converter in the Y – direction. Thus k_y/k_x is taken to be 500. So L_a found to be 131.75μm and W_{sp} equals to 53.95μm. The output power of the converter is small (12.7μW).

V. INCREASING THE OUTPUT POWER BY INCREASING THE CONVERTER THICKNESS

From equation (9), to increase P_{out} one has to decrease the factor α. This can be achieved by increasing C_{max} for a certain value of C_{par}. By increasing the device thickness (t), C_{max} increases. As a case study, let the device thickness (t) be 500μm [3]. Therefore C_{max} will be 1480pF. Thus P_{out} will be 82.8μW. It is obvious that, by increasing the converter thickness, the output power of the converter can be enhanced.

VI. CONCLUSION

The equation of the output power from the converter including the effect of C_{par} is presented. The output power of the converter is 12.7μW if the converter thickness is 25μm. By increasing the converter thickness to 500μm the output power becomes 82.8μW.

REFERENCES

[1] S. Roundy, P. K. Wright and K. S. J. Pister, David A. Dornfeld "Energy Scavenging for Wireless Sensor Nodes with a Focus on Vibration to Electricity Conversion", A dissertation of Philosophy in Engineering-Mechanical Engineering in the UNIVERSITY OF CALIFORNIA, BERKELEY, Spring 2003.
[2] S. Meninger, "A Low Power Controller for a MEMS Based Energy Converter", Master of Science at the Massachusetts Institute of Technology, 1999.
[3] Scott Meninger, Jose Oscar Mur-Miranda, Rajeevan Amirtharajah, Anantha P. Chandrakasan, and Jeffrey H. Lang,, "Vibration-to-Electric Energy Conversion", IEEE TRANSACTIONS ON VERY LARGE SCALE INTEGRATION (VLSI) SYSTEMS, VOL. 9, NO. 1, FEBRUARY 2001
[4] http://www.memscap.com/memsrus/crmumps.html
[5] Gary Keith Fedder, "Simulation of Micro-electromechanical Systems", a dissertation of Philosophy in Engineering-Mechanical Engineering in the UNIVERSITY OF CALIFORNIA, BERKELEY, 1994.
[6] http://www.Silicon Properties.html.

Highly Efficient Micromachined Bragg Mirrors Using Advanced DRIE Process

B. SAADANY, D. KHALIL, and T. BOUROUINA

Abstract— A novel advanced Deep Reactive Ion Etching (DRIE) process technique is used to realize highly efficient Vertical Bragg mirrors. The Bragg mirrors are realized by anisotropic etching of Si using DRIE, thus producing successive vertical interfaces between Si and air. The new etching technique is based on combining 2 steps of Cryogenic and Bosch DRIE processes to obtain high quality Si walls in terms of both: high aspect ratio vertical walls as well as smooth surface. The realized Bragg mirrors, fabrication process, as well as measured optical performance showing the advantages of the new technique are presented.

Index Terms— DRIE, Bragg Mirrors, Cryogenic DRIE

I. INTRODUCTION

Bragg mirrors have numerous applications when optical filtering is concerned, which is essential for current as well as next generation high speed optical communication systems and networks. Optical filters represent the fundamental part of the Optical Add Drop Multiplexer (OADM), which is an essential element for all DWDM systems [1]. They are also required for channel monitoring [2], dispersion equalization [3] as well as improving the spectral purity and the tunability of the optical source. They are also indispensable components in the field of environmental sensing and spectrometric applications [4]. Bragg mirrors integrated in MEMS devices are attractive candidates to such applications as they have all the MEMS advantages of batch processing, low cost, and compatibility with standard microelectronics. Moreover, MEMS technology with its numerous actuation techniques enables the realization of new functions and features of photonic devices such as optical tunability as well as numerous dynamic sensing applications. The proposed micro-machined Bragg mirrors have thus been used to realize MEMS photonic devices such as Tunable Optical Filters [5-10], Optical ADD/Drop multiplexers [11], as well as the recently reported Michelson Interferometer used as a spectrometer [12]. Performance of all these photonic devices proved to be highly sensitive to the quality of the realized Bragg mirrors.

In this work, we propose a novel Deep Reactive Ion Etching (DRIE) process technique to realize highly efficient micro-machined Bragg mirrors for optical MEMS devices.

First, the principle of operation of the micromachined Bragg mirrors is explained, followed by a presentation of the fabrication process steps as well as the enhanced fabrication process using the proposed DRIE technique. Finally, experimental measured results showing the advantage of the new technique are discussed.

II. PRINCIPLE OF OPERATION

The proposed micromachined Bragg mirrors are realized by anisotropic etching of small trenches in the silicon substrate, thus realizing vertical Bragg layers. The width of each Bragg layer is first defined by lithography, then anisotropic etching is performed using Deep Reactive Ion Etching (DRIE). Thus refractive index modulation is obtained by the successive vertical interfaces between Si (high n) and Air (low n). We thus consider *free-space propagation* of light *in the plane of the substrate*. This approach has several advantages compared to conventional methods for the realization of Bragg mirrors, where deposition of multilayer stacks of alternating dielectric thin films is usually used, and thus light propagation in such conventional Bragg mirrors is perpendicular to the plane of the substrate. The schematic diagram shown in Fig. 1 below illustrates the difference between the proposed approach and the conventional approach for realizing Bragg mirrors.

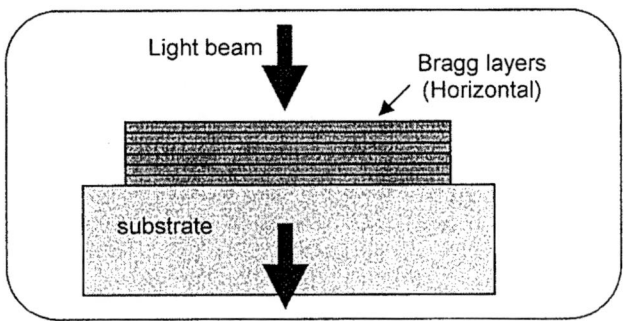

a) Schematic diagram of a conventional "horizontal" Bragg mirror

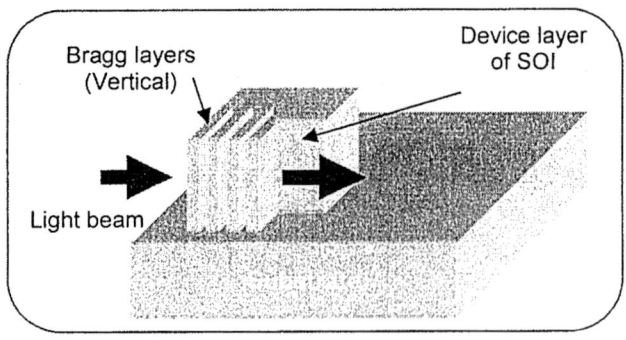

b) The proposed micromachined "vertical" Bragg mirror

B. Saadany and T. Bourouina are with ESIEE/Esycom-Lab, Cité Descartes 2 Bd Blaise Pascal, 93162 Noisy-le-Grand Cedex FRANCE Phone (33) 1 45926692 Fax (33) 1 45926699, email (saadanba@esiee.fr, and t.bourouina@esiee.fr)

D. Khalil is with Ain-Shams University, 1 El-Sarayat St., Abassia, Cairo, Egypt (e-mail: diaa.khalil@ieee.org).

1-4244-0899-7/06/$20.00 ©2006 IEEE.

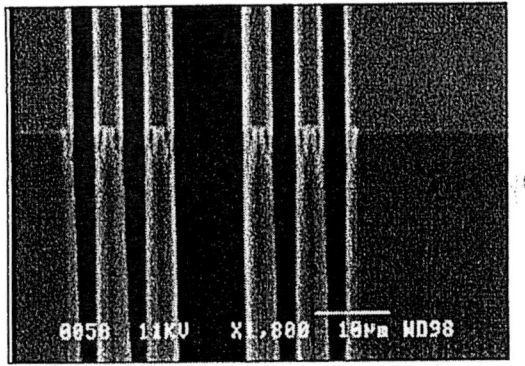

c) An SEM of a realized micromachined "vertical" Bragg mirror

Fig. 1 The principle of operation of the micromachined vertical Bragg mirrors

The main advantages of the proposed vertical Bragg mirrors can be summarized as follows:

1- Bragg mirrors are vertical and can be movable using MEMS actuators when realized over SOI, where Bragg etching would then be on the device layer of the SOI wafer. Optical MEMS devices such as Optical filters [5-11] and actuated Michelson interferometers [12] were realized using this technique.

2- Bragg layers are defined by Lithography, thus allowing for flexibility in the mirror design with no restrictions on tunability or mirror curvature. For example, cylindrical cavity, coupled cavity or cascaded Fabry-Pérot filters can all be realized and tuned much easier than with conventional techniques.

3- Fiber grooves as well as device actuators (when required) are all defined using the same lithography step, thus allowing for very good alignment and easier assembly. This naturally falls in the side of better performance as well as overall device cost reduction.

Technological constraints on the other hand are certainly a challenge. Fabrication of vertical Bragg mirrors requires etching of narrow and deep features. They must be narrow enough to be in the order of a few quarter wavelengths and deep enough to facilitate coupling of light with optical fibers. Thickness resolution of the realized Bragg mirrors requires an accurate estimation of lithographic as well as etching induced errors, where such errors need to be accounted for in the Bragg mirror design. Highly vertical etching as well as a final smooth surface of the etched Si walls is essential for obtaining a high quality Bragg mirror. This was one of our motivations to develop the proposed high aspect ratio smooth surface DRIE process technique.

III. FABRICATION PROCESS

The fabrication process steps are shown in Fig. 2, where as shown an Aluminum sputtering step is used to form an Aluminum layer that would function as a mask for DRIE. One lithography step is sufficient for realizing the Bragg layers, fiber grooves, as well as the electrostatic actuator. Even with the loading effect of DRIE (etching depth dependence on feature size) which can cause the etching

depth of Bragg trenches not to reach the SOI etch stop layer. However, since light is coupled to the filter through optical fibers (diameter 125 µm), thus the illuminated area of the filter is mainly on the upper part of the Bragg layers –see Fig. 3(a)-. However, in the case of the moving Bragg mirrors such as tunable filter, care must be taken for the resonator gap feature size to be large enough for DRIE to reach the SiO_2 layer of the SOI, thus enabling the release of the moving Bragg mirror.

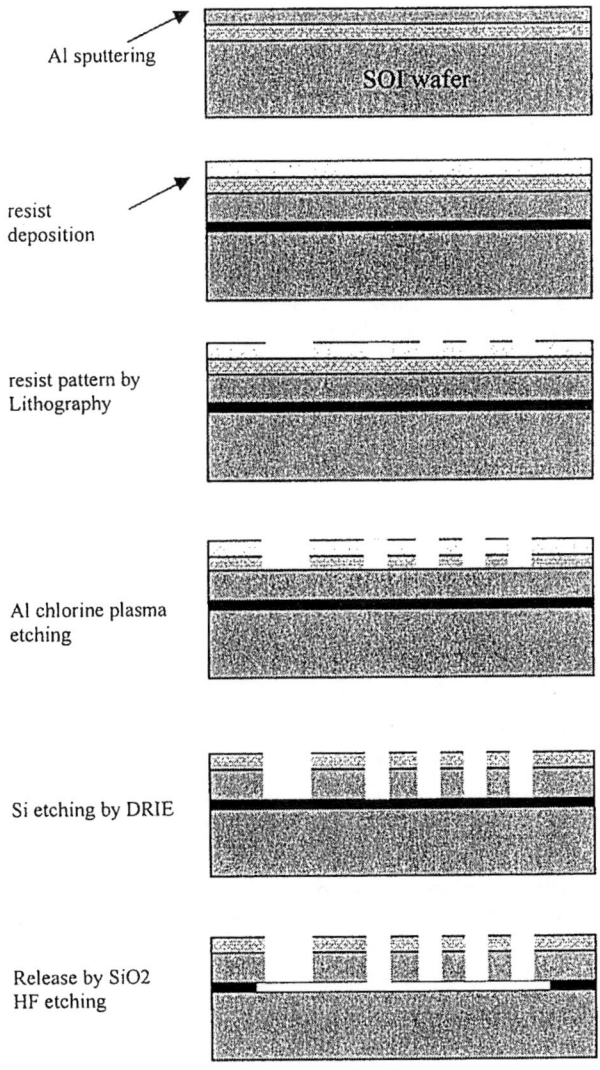

Fig.2 . Fabrication process steps.

The lithography pattern is transferred to the Aluminum mask by plasma chlorine etching of the Aluminum layer. DRIE is then applied to the wafer until reaching the oxide etch stop layer of the SOI. Then the structure is released by etching the buried SiO_2 layer using HF. In the case of using the *Bosch* process [13] for DRIE, where a rough surface of the Bragg mirrors is obtained due to the periodic nature of the process. Growing a thin oxide layer on Bragg surfaces may be useful before structure release to ensure a smoother Bragg surface. It is worth mentioning here that if cryogenic DRIE process is used, this step would be unnecessary since smoother surface is obtained directly after etching. Also, the Aluminum mask will not be necessary, as the process tends to have better selectivity between photoresist and Si at lower temperatures. However, cryogenic DRIE can be optimized for wall verticality for Bragg layers with small feature sizes, but may not be an optimum process for the large feature sizes (e.g. electrostatic actuator). Thus, a more robust design for electrostatic actuator is needed in cryogenic DRIE process.

A. Enhanced process (2-step: Cryogenic/Bosch processes)

Although the Bosch DRIE process is relatively stable and enables the realization of different feature sizes with good verticality, its main disadvantage is the surface roughness of the resulting surfaces, (Bragg mirrors in our case), which in turn degrades the overall filter characteristics. This roughness results from the cyclic nature of the process. Cryogenic DRIE process on the other hand produces smooth surfaces. However, the obtained verticality is more sensitive to the etched feature sizes. Also, for deep structures the undercut effect (although a small percentage of the feature size), is more prominent in this case due to the sensitivity of the Bragg layers to width resolution.

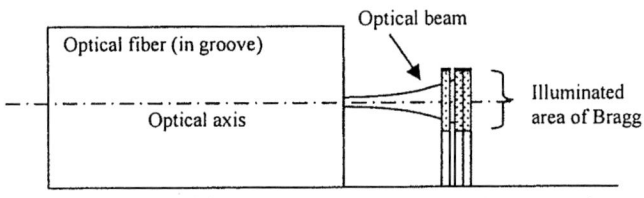

a) Cross section sketch of optical illumination of the vertical Bragg mirror

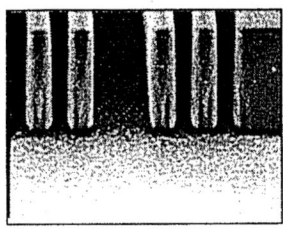

b) SEM after DRIE Step1 :
cryogenic process (smooth surface)

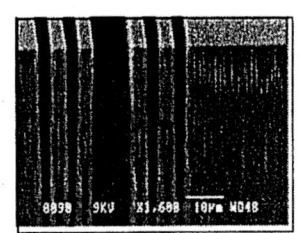

c) SEM after DRIE Step2 :
Bosch process

Fig. 3 Enhanced 2-step etching process

In order to obtain the advantages of both processes (Bosch & cryogenic), an enhanced process that utilizes both processes is presented. The process takes benefit from the fact that the illuminated area of the vertical Bragg mirrors is limited to the upper part of the mirror as shown in Fig. 3 (a). Thus, cryogenic DRIE is applied first, until reaching an etching depth that is enough to cover the illuminated area of the Bragg mirror (about 30 μm). Then a passivation step of plasma C4F8 similar to that of Bosch process is applied to the etched walls for protection. Then etching is completed by Bosch DRIE process, thus better maintaining the etching verticality for the different feature sizes of the design. Fig. 3(b) and (c) show SEM photos of the two consecutive steps of DRIE etching of a realized Bragg mirror.

IV. MEASURED RESULTS

For measuring the Bragg mirrors spectral response, a tunable laser source (range 1524 – 1576 nm) is used to scan the different wavelengths, where output power is detected, thus obtaining the filter response. The tunable laser output is coupled with visible red laser to facilitate alignment and visualization, where optical fibers are used for injecting the tunable laser beam to the sample, as well as for detecting the output power.

Different Bragg mirror filters processed using different process conditions were measured. In the following, some selected characteristics are presented to illustrate the effect of such process conditions. For example, transmission responses of two filters composed of 4 Si wall Bragg mirrors processed using normal Bosch DRIE are shown in Fig. 4. In Fig. 4(a) the measured response of the first transmission filter results in a relatively wide FWHM of about 6 nm, while as shown in Fig. 4(b) there are some side lobes near the resonance wavelength of the notch filter response. In comparison, less than 0.4 nm FWHM is obtained with smooth Bragg surfaces compared to about 2 nm FWHM for the notch filter of Fig. 4(b) (unsmoothed Bragg surface).

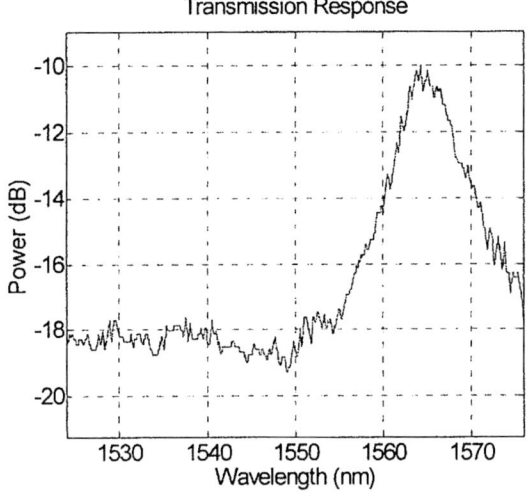

a) Transmission filter resonance wavelength at 1564.25 nm

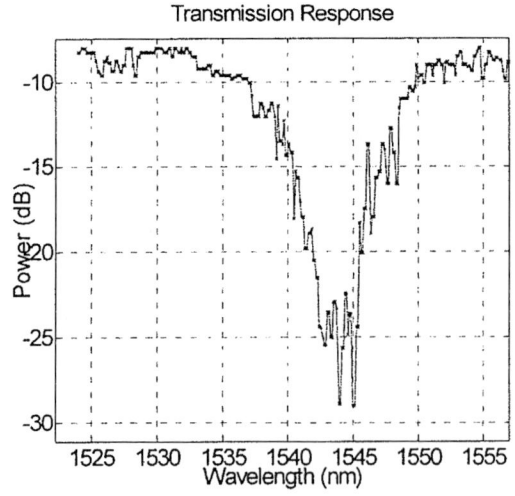

a) Notch filter resonance wavelength at 1544.5 nm

Fig. 4 Measured Transmission spectrum of two different filters realized using normal Bosch process

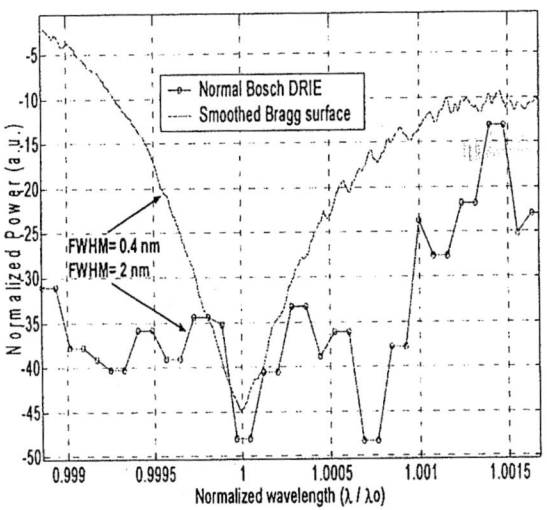

Fig. 5 Effect of smooth Bragg surface on measured filter selectivity

The effect of the proposed 2-step DRIE technique can be clearly detected in Fig.5, where the two compared filter characteristics are shown in a normalized scale to illustrate such effect. We can clearly notice the enhanced selectivity of the smooth Bragg surface, emphasized by the 5X (0.4:2 nm) ratio enhancement of the FWHM. Finally, at Fig. 6 we can see an example of the Fabry-Pérot cavity [10] with multiple resonance around the 1550 nm wavelength, where the proposed 2-step DRIE process was again used to realize the whole device including Bragg mirrors, fiber grooves, as well as the actuator. We can see here that the FWHM is almost constant (FWHM = 1.2 nm) for the different resonance wavelengths.

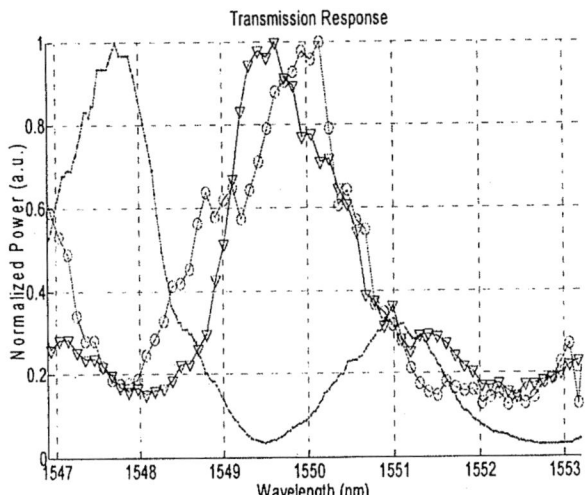

Fig. 6 Measured multiple resonance around 1550 nm wavelength '

V. CONCLUSION

We have demonstrated the realization of micromachined Bragg mirrors using an enhanced 2-step DRIE etching of Si to obtain high quality vertical Bragg mirrors. The 2-step (Cryo + Bosch) DRIE showed a performance advantage of the realized Bragg mirror filters, where a 5X FWHM selectivity enhancement was noticed. The technique was also used to realize a complete tunable filter device with a

stable 1.2 nm FWHM for different resonance wavelengths around 1550 nm.

VI. ACKNOWLEDGEMENT

This project had partial support from Conseil Regional d'Ile de France, EGIDE France and the Japan Society for Promotion of Science (JSPS). The silicon devices presented in this paper were realized in the SMM clean room facilities of ESIEE. E-beam lithography was performed in VDEC at The University of Tokyo. Optical measurements were performed at the Laser lab, of Ain-Shams University, Cairo.

REFERENCES

[1] D. Sadot, E. Boimovich, "Tunable Optical Filters for Dense WDM Networks", IEEE Comm. Mag. , Vol 36, Dec 1998
[2] J.H. Chen, Y. Chai, J.Y. Fan, F.S.Choa, T. Tanbun, R. Logan, W. Tsang, C. Burrus, " WDM channel monitoring and signa; power control/equalization using integrated tunable active filters", IEEE/LEOS summer Top. Meeting 1997
[3] Y. Song, D. Starodubov, Z. Pan, Y. Xie, A. Willner, J. Feinberg, " A tunable dispersion compensator with fixed bandwidth for WDM systems using a uniform FBG", CLEO 2001.
[4] G. Lammel, S. Schweizer, P. Renaud, "MEMS infrared gas spectrometer based on a porous silicon tunable filter", Micro Electro Mechanical Systems, 2001. MEMS 2001. The 14th IEEE International Conference on, 2001.
[5] J.H. Lee, S. Yun « Micromachined in-plane tunable optical filter using thermo-optic effect ». Proceedings of SPIE Vol. #4983 (2003) 195-202
[6] B. Saadany, T. Bourouina, D. Khalil, "Design of a MEMS Tunable Optical Filter Based on a Novel Vertical DBR Architecture", MEMSWAVE Conference, Toulouse, France, July 2-4 (2003) Poster.
[7] S.S. Yun, K.W. Jo and J.H. Lee « Crystalline Si-based In-plane Tunable Fabry-Perot Filter with Wide Tunable Range » IEEE/LEOS Optical MEMS and their applications, Hawaii, August 19 (2003)
[8] S.S. Yun and J.H. Lee "A micromachined in-plane tunable optical filter using the thermo-optic effect of crystalline silicon". J. Micromech. Microeng. 13 (2003) 721-725
[9] B. Saadany, F. Marty, Y. Mita, D. Khalil and T. Bourouina, 'A MEMS Tunable Optical Filter Based on Vertical DBR Architecture'. DTIP'04; Design, Test, Integration and Packaging of MEMS and MOEMS, Montreux, Switzerland, May 12-14 (2004).
[10] B. Saadany, M. Malak, F. Marty, Y. Mita, D. Khalil, and T. Bourouina, "Electrostatically-tuned Optical Filter Based on Silicon Bragg Reflectors", Optical MEMS 2006, Big Sky, MT, USA, August 21-24 (2006).
[11] B. Saadany, D. Khalil, M. Malak, M. Kubota, F. Marty, Y. Mita, and T. Bourouina, "An all Silicon Micro-machined Add-Drop Optical Filter", Optical MEMS 2006, Big Sky, MT, USA, August 21-24 (2006).
[12] B. Saadany, T. Bourouina, M. Malak, M. Kubota, Y. Mita, and D. Khalil, "A Miniature Michelson Interferometer using Vertical Bragg Mirrors on SOI", Optical MEMS 2006, Big Sky, MT, USA, August 21-24 (2006).
[13] Patents DE4241045, US 5501893 and EP 625285, authors Franz Lärmer, Andrea Schilp.

Monte Carlo Simulation of Photonic Band Gap Structures

Tarek Badreldin and Diaa Khalil, *Senior Member, IEEE*

Abstract— We developed a statistical Monte Carlo technique for the performance analysis of Photonic band gap structures. The randomness nature of the fabrication process of the photonic crystals is taken into account in this analysis. The technique is applied on the bandgap calculation of two-dimensional photonic crystals to study the effect of manufacturing imperfections on the photonic band gap. This helps to establish a design for manufacturability for the photonic crystals.

Keywords— Photonic Crystals, Monte Carlo simulation, 2D photonic crystals, process capability, design for manufacturability, tolerance.

I. INTRODUCTION

PHOTONIC Band gap structures (also called photonic crystals) are nowadays one of the main research directions in integrated optics as they represent a technology platform that can be used for the fabrication of many compact integrated passive and active optical components [1-4]. Their unique properties in light guiding and control have attracted the attention of many research groups and thus a huge number of publications are currently published in this area each month. However, the industrialization of photonic band gap structures is still very limited. This is mainly due to the high technological barrier required to obtain reliable reproducible photonic band gap structures fabricated with a reasonable yield. The fact that these structures require dimension control in the order of few nanometers in all directions explains this difficulty. The quantitative evaluation of this difficulty is thus of great importance for the future development of photonic crystals applications. This could be done by evaluating the effect of the process parameters (technology limitations) on the main feature of the photonic crystals, that is, the photonic band gap. This is the objective of the work presented in this paper.

To allow for process parameters evaluations, we developed a Monte Carlo simulation technique for the evaluation of the optical MEMS components [5]. This technique has been successively applied on the case of optical MEMS switch. In this work we develop a similar technique for the photonic band gap structures to evaluate the effect of the variations in process parameters on the band gap of a 2D photonic

crystal. As the photonic band gap is not calculated by a simple analytical expression as in the simple case of the 2x2 optical MEMS switch, the plan wave expansion method is developed with a built in Monte Carlo technique based on random number generation associated to each process parameters. The theoretical base of the plan wave expansion method will be detailed in section II while its numerical application will be explained in section III. The development of the Monte Carlo technique is presented in section IV and the obtained results are discussed in section V.

II. BACKGROUND: 2D PHOTONIC CRYSTALS ANALYSIS

Starting from Maxwell's equations, assuming ideal lossless medium with no current sources, we have [6]:

$$\nabla \cdot \bar{D}(r,t) = \varepsilon_o \nabla \cdot \left[\varepsilon_r(r)\bar{E}(r,t) \right] = 0 \qquad (1)$$

$$\nabla \cdot \bar{B}(r,t) = \mu_o \nabla \cdot \bar{H}(r,t) = 0 \qquad (2)$$

$$\nabla \times \bar{E}(r,t) = -\frac{\partial \bar{B}(r,t)}{\partial t} = -\mu_o \frac{\partial \bar{H}(r,t)}{\partial t} \qquad (3)$$

$$\nabla \times \bar{H}(r,t) = \frac{\partial \bar{D}(r,t)}{\partial t} = \varepsilon_o \varepsilon_r(r) \frac{\partial \bar{E}(r,t)}{\partial t} \qquad (4)$$

where the relative magnetic permeability is assumed equal to unity, i.e.

$$\mu_r(r) = 1 \qquad (5)$$

For two-dimensional photonic crystal, the dielectric structure is uniform in the z direction as shown in fig. (1). The electromagnetic waves travel in the x-y plane and are also uniform in the z direction. Hence, ε(r), E(r) and H(r) are independent of the z-coordinate in (3) and (4). In this case, these vectorial equations are decoupled to two independent sets of equations. The first set is

$$\frac{\partial}{\partial y} E_z(r_{//},t) = -\mu_o \frac{\partial}{\partial t} H_x(r_{//},t) \qquad (6)$$

$$\frac{\partial}{\partial x} E_z(r_{//},t) = \mu_o \frac{\partial}{\partial t} H_y(r_{//},t) \qquad (7)$$

$$\frac{\partial}{\partial x} H_y(r_{//},t) - \frac{\partial}{\partial y} H_x(r_{//},t) = \varepsilon_o \varepsilon(r_{//}) \frac{\partial}{\partial t} E_z(r_{//},t)$$

$$(8)$$

And the second set is

Tarek Badreldin is with Mentor Graphics, Egypt. (e-mail: tarek_badreldin@mentor.com)
Diaa Khalil is with the faculty of Engineering, Ain Shams University, 1 El-Sarayat Street, Abbassia, Cairo, Egypt. (e-mail: diaa.Khalil@ieee.org)

$$\frac{\partial}{\partial y} H_z(r_{//},t) = \varepsilon_o \varepsilon(r_{//}) \frac{\partial}{\partial t} E_x(r_{//},t) \qquad (9)$$

$$\frac{\partial}{\partial x} H_z(r_{//},t) = -\varepsilon_o \varepsilon(r_{//}) \frac{\partial}{\partial t} E_y(r_{//},t) \qquad (10)$$

$$\frac{\partial}{\partial x} E_y(r_{//},t) - \frac{\partial}{\partial y} E_x(r_{//},t) = -\mu_o \frac{\partial}{\partial t} H_z(r_{//},t) \quad (11)$$

where

$$\vec{r}_{//} = x\,\hat{a}_x + y\,\hat{a}_y \qquad (12)$$

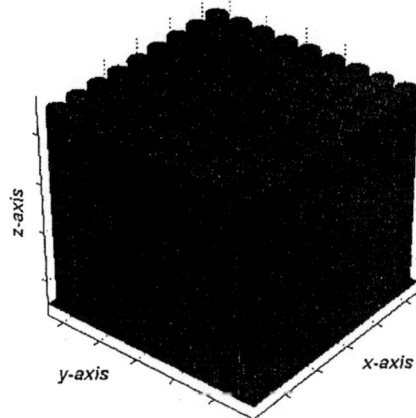

Fig. 1 Two-dimensional photonic crystals structure with rods in air background

Eliminating $H_x(r_{//},t)$ and $H_y(r_{//},t)$ from the first set, we obtain the wave equation for $E_z(r_{//},t)$

$$\frac{1}{\varepsilon(r_{//})}\left\{\frac{\partial^2}{\partial x^2}+\frac{\partial^2}{\partial y^2}\right\}E_z(r_{//},t)=\frac{1}{c^2}\frac{\partial^2}{\partial t^2}E_z(r_{//},t) \quad (13)$$

From the second set, we obtain the wave equation for $H_z(r_{//},t)$

$$\left\{\frac{\partial}{\partial x}\frac{1}{\varepsilon(r_{//})}\frac{\partial}{\partial x}+\frac{\partial}{\partial y}\frac{1}{\varepsilon(r_{//})}\frac{\partial}{\partial y}\right\}H_z(r_{//},t)=\frac{1}{c^2}\frac{\partial^2}{\partial t^2}H_z(r_{//},t) \quad (14)$$

We seek a solution in the form of

$$E_z(r_{//},t) = E_z(r_{//})e^{-iwt} \qquad (15)$$

$$H_z(r_{//},t) = H_z(r_{//})e^{-iwt} \qquad (16)$$

The eigenvalue equations are thus given by

$$-\frac{1}{\varepsilon(r_{//})}\left\{\frac{\partial^2}{\partial x^2}+\frac{\partial^2}{\partial y^2}\right\}E_z(r_{//})=\frac{\omega^2}{c^2}E_z(r_{//}) \quad (17)$$

$$-\left\{\frac{\partial}{\partial x}\frac{1}{\varepsilon(r_{//})}\frac{\partial}{\partial x}+\frac{\partial}{\partial y}\frac{1}{\varepsilon(r_{//})}\frac{\partial}{\partial y}\right\}H_z(r_{//})=\frac{\omega^2}{c^2}H_z(r_{//}) \quad (18)$$

These two kinds of eigenfunctions represent two independent polarizations; the E-polarization for which the electric field is parallel to the z-axis, and the H-polarization for which the magnetic field is parallel to the z-axis.

Now we apply the Bloch's theorem to express $E_z(r_{//})$ and $H_z(r_{//})$ as follows

$$E_z(r_{//}) = E_{z,k_{//}n}(r_{//}) = \sum_{G_{//}} E_{z,k_{//}n}(G_{//})e^{i(\vec{k}_{//}+\vec{G}_{//}).\vec{r}_{//}} \quad (19)$$

$$H_z(r_{//}) = H_{z,k_{//}n}(r_{//}) = \sum_{G_{//}} H_{z,k_{//}n}(G_{//})e^{i(\vec{k}_{//}+\vec{G}_{//}).\vec{r}_{//}} \quad (20)$$

where

$$\vec{G}_{//} = \frac{2\pi}{a_o}(l_1 a_x + l_2 a_y) \qquad (21)$$

l_1 and l_2 are arbitrary integers, and a_o is the lattice constant.

Substituting with (19) and (20) into (17) and (18) respectively, we obtain the following eigenvalue equations for the expansion coefficients

For the E-polarization

$$\sum_{G'_{//}}\kappa(G_{//}-G'_{//})\left|\vec{k}_{//}+\vec{G}'\right|^2 \vec{E}_{z,k_{//}n}(G'_{//}) =$$

$$\frac{\omega^{(E)}_{k_{//}n}{}^2}{c^2}\vec{E}_{z,k_{//}n}(G_{//}) \qquad (22)$$

while for the H-polarization

$$\sum_{G'_{//}}\kappa(G_{//}-G'_{//})(\vec{k}_{//}+\vec{G}')\cdot(\vec{k}_{//}+\vec{G})\vec{H}_{z,k_{//}n}(G'_{//}) =$$

$$\frac{\omega^{(H)}_{k_{//}n}{}^2}{c^2}\vec{H}_{z,k_{//}n}(G_{//}) \qquad (23)$$

where κ is the reciprocal lattice constant given by

$$\frac{1}{\varepsilon(r)} = \sum_{G}\kappa(G)e^{-iG.r} \qquad (24)$$

III. Numerical Analysis

We will focus on the E-polarization given by (22) as an example and describe each of the components of the equation briefly and show how to calculate it numerically.

1- Lattice Properties:
The lattice properties are the lattice geometry (lattice geometry, lattice constant a_o and rods radius r_a) and the lattice foreground and background refractive indices.

2- The wave vector $\vec{k}_{//}$:

It consists of 2 vectors parallel to the substrate, $\vec{k}_x$ and $\vec{k}_y$.

Each component varies from 0 to π/a_o with step δk

$$\vec{k}_{//} = k_x + k_y \qquad (25)$$

3- The reciprocal refractive index

$$\kappa(G_{//} - G'_{//}) = \frac{1}{V_o} \int_{V_o} \frac{1}{\varepsilon(r_{//})} e^{-i(\vec{G}_{//} - \vec{G}'_{//}).\vec{r}} dr \qquad (26)$$

This integration could be solved to get an expression for $\kappa(G_{//} - G'_{//})$ in the case of periodic, defect-free photonic crystal structure, where ε(r) can have a regular expression. However, when a defect exists in the lattice, numerical integration could be done over a "unit cell" of the crystal, using the Fourier Expansion (26), assuming that the unit cell is repeated periodically over the whole crystal structure. In our case, we consider imperfections randomly generated in the crystal geometry and thus the unit cell approximation will be a more suitable solution.

IV. MONTE CARLO SIMULATION

Our Monte Carlo Engine implementation is mainly based on the use of random number generators that assign, at each process, values for the different random parameters used in the calculations. For each of these parameters, a probability density function PDF, with a specific average and standard deviation, is assigned. In our analysis, the random variables considered in the structure geometry are

- The rods radius (R_x and R_y)
- The rods center position (X and Y)

Each of these 4 parameters affects the value of ε(r) in (26), and thus the value of the reciprocal refractive index integration.

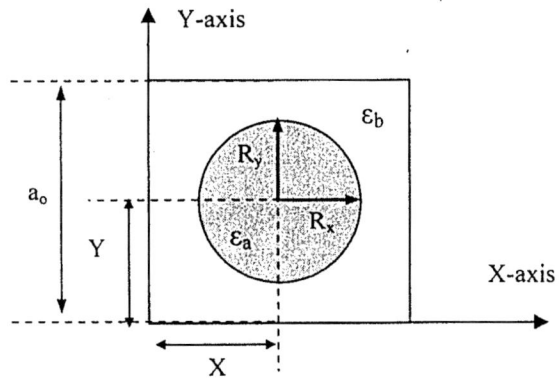

Fig. 2 Photonic crystal parameters used in the analysis as random variables

Variations in the material refractive index are not studied for the time being. Each one of the 4 random variables (R_x, R_y, X and Y) is assumed to be completely independent of the others and to have a truncated Gaussian probability density function similar to that shown in Fig. (3).

Fig. (4) shows an ideal 9x9 photonic crystal structure without any manufacturing imperfections. The corresponding band structure is shown in Fig. (5). When introducing the randomness on the variables X, Y, R_x and R_y, the structure takes the form shown in Fig. (6)., where we can easily note the deviations from the ideal structure. Such variations represent the random errors that come from the limits in the resolution of the fabrication technology either the lithography or the etching process. The probability density function of the variation in both the rod positions and radii is assumed to have the Gaussian distribution

shown in Fig. (3) with a standard deviation of 10% of both the lattice constant and the rod radii. The higher and lower limits of the variation are also assumed to be +/- 10% of the lattice constant a_o for the position, and of the radius r_a for the radii variation. We call such a case a 10% tolerance or error. The ratio of the lattice constant to the rod radius is 1:0.2. The foreground refractive index ε_a is assumed to be 9 while the background ε_b is assumed to be 1 (air). Fig. 7 shows the band diagram corresponding to this case of 10% tolerance. It is clear, when we compare figures (5) and (7), that the random variations in the rod positions and radii cause a narrowing in the structure band gap of 35%.

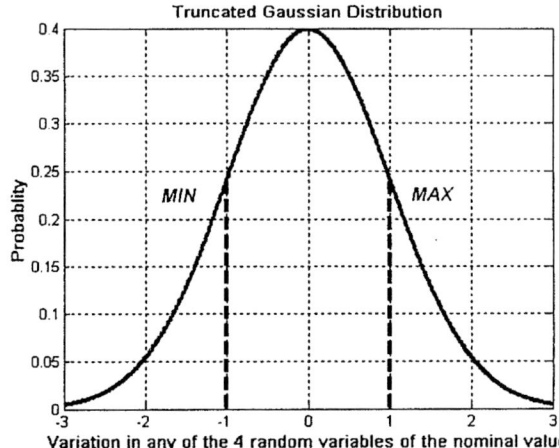

Fig. 3 Truncated Gaussian Probability density function used in the generation of each of the 4 random variables (R_x, R_y, X and Y)

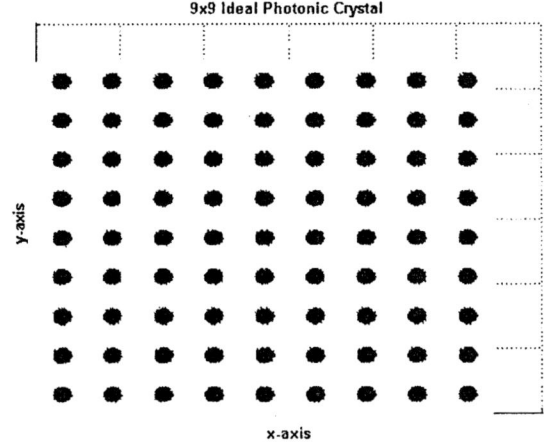

Fig. 4 Ideal 9x9 Square-lattice Photonic Crystal Structure, a_o: r_a = 1:0.2

V. APPLICATION: BANDGAP CALCULATION

As shown in section IV, the defects in the rod radius or position, modeled through the Monte Carlo simulation, cause a variation in the crystal bandgap. We can extent the calculations to show how the bandgap varies with the manufacturing tolerance or process variation δ. Figure (8) shows the variation of the band gap of the 9x9 photonic

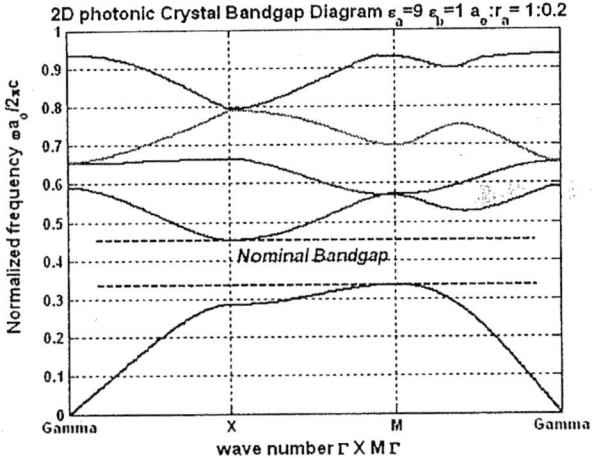

Fig. 5 Normalized Frequency Bandgap diagram of Ideal 9x9 Photonic Crystal Structure

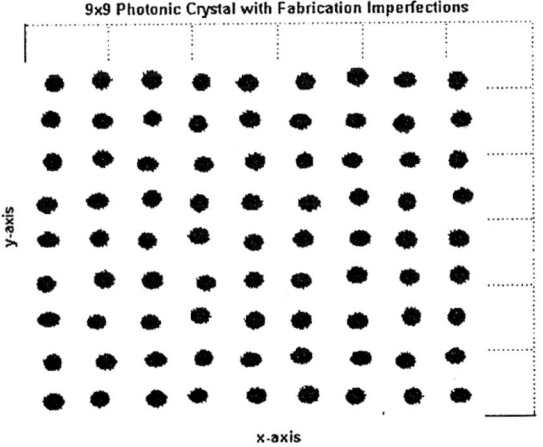

Fig. 6 Non-Ideal 9x9 Square-lattice Photonic Crystal Structure with error standard deviation and limit = 10%, $a_o: r_a = 1:0.2$

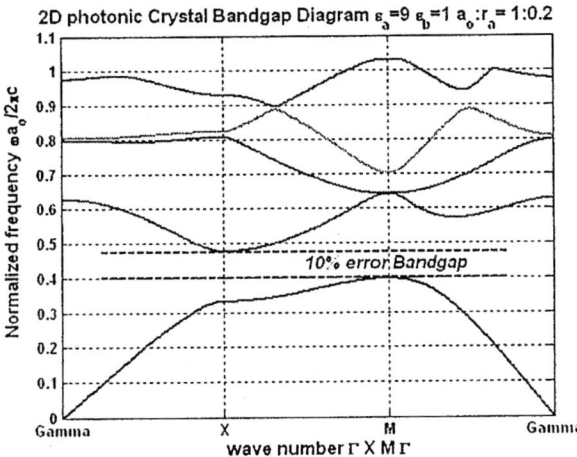

Fig. 7 Normalized Frequency Bandgap diagram of Non-Ideal 9x9 Photonic Crystal Structure in the case of 10% tolerance (error) value.

crystal shown in Fig. (4) as a function of the manufacturing tolerance or resolution. It is calculated as:

$$Bandgap = Root_2(X) - Root_1(M) \quad (27)$$

$$X = wave\_number\,(k_x = \frac{\pi}{a_o}, k_y = 0) \quad (28)$$

$$M = wave\_number\,(k_x = \frac{\pi}{a_o}, k_y = \frac{\pi}{a_o}) \quad (29)$$

where X and M are the wave numbers shown in Fig. (5) x-axis. We can see that at $\delta = 0\%$, the nominal normalized bandgap value is 0.118. As the uncertainty in the process parameters increases, the bandgap value decreases. When δ value reaches 0.17 or 17%, the bandgap is completely closed. After this value, there is no bandgap existing and the value calculated represents the difference between the second and first bands. When using a 9x9 structure (as that one shown in Fig. (6)), it is assumed that, at a certain tolerance value, this structure is periodically repeated in both the x and y directions, with the exact rods positions and radii. Figure (9) shows the bandgap percentage variation relative to the nominal bandgap value, 100% is the ideal case with no variations in rods radii or positions, 0% is complete bandgap closure. The analysis is done for 3x3, 5x5 and 9x9 photonic crystals structures. We can see that although the 3 structures give very close results, the results get more scattered for small structures, i.e. the 3x3 data results are more scattered (i.e. more sensitive to the process tolerance) than the results of the 9x9 structure.

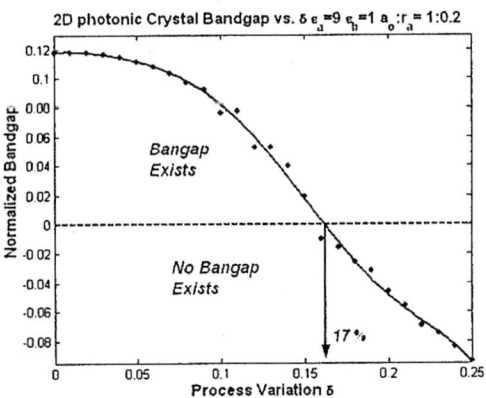

Fig. 8 The variation of the normalized bandgap of 9X9 Photonic Crystal vs. the Process Variation δ. (Dotted: Actual Simulation Data, Solid line: Curve Fitting)

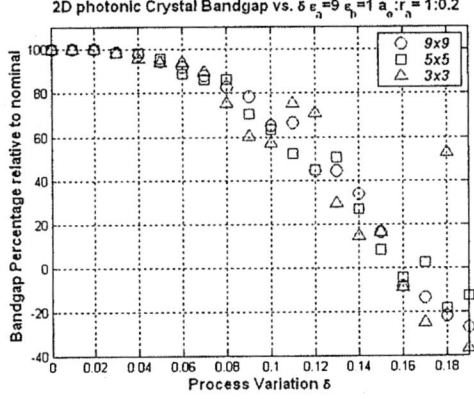

Fig. 9 The percentage variation of the bandgap of different Photonic Crystal Structures vs. the Process Variation δ. (Dots: 9x9, Squares: 5x5 and Triangles: 3x3)

These results show that we can use this technique to identify the process yield with respect to a certain tolerance value (δ), through the calculation of process capability parameters. To achieve this target, we need to carry out the bandgap calculation for a certain tolerance value for a number of samples. Let us consider the case of a process tolerance of 10% for both the rods radius and position. Figure (10-12) shows the sample results for the band gap calculation for different sizes of photonic crystals plotted vs. the variation in the x-position of the rods. Figures (13-15) show the same bandgap samples plotted vs. the rods x-axis radius (R_x) variation. For the y-position and y-axis radius (R_y), the figures will look very similar. We can see that for all sizes, most of the samples fall between 60% and 80% of the nominal, ideal bandgap value.

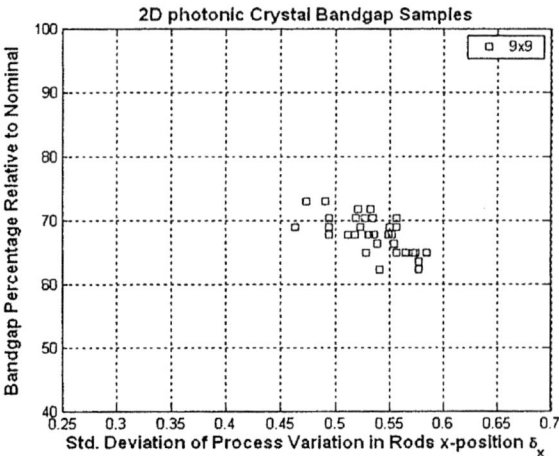

Fig. (12) Bandgap calculation samples of Square-lattice Photonic Crystal structure of 9x9 structure plotted vs. the rods variation off the nominal position in the x-axis.

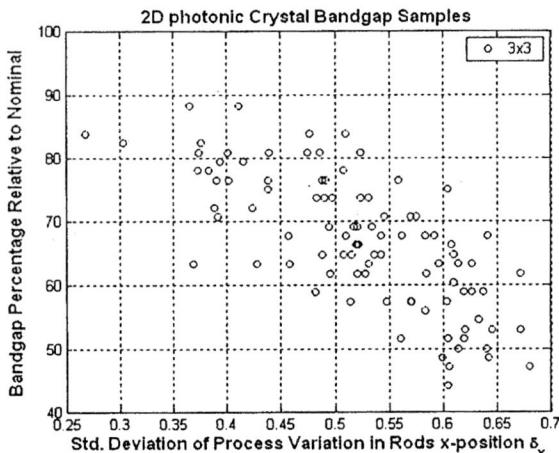

Fig. (10) Bandgap calculation samples of Square-lattice Photonic Crystal structure of 3x3 structure plotted vs. the rods variation off the nominal position in the x-axis.

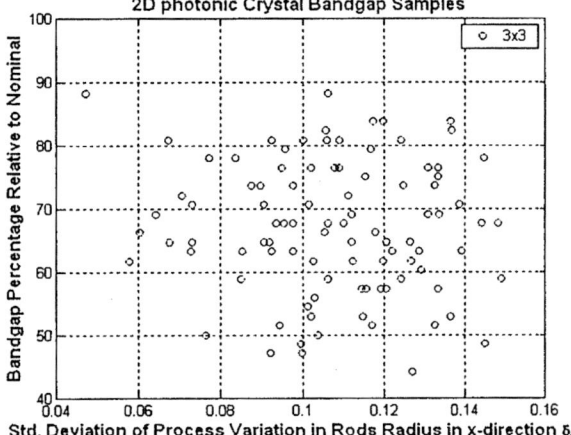

Fig. (13) Bandgap calculation samples of Square-lattice Photonic Crystal structure of 3x3 structure plotted vs. the rods radius variation in the x-axis.

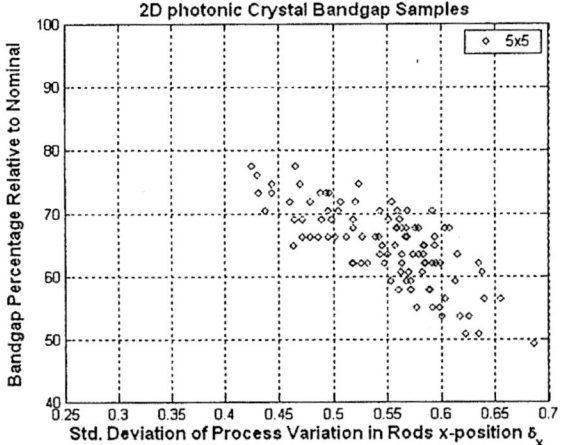

Fig. (11) Bandgap calculation samples of Square-lattice Photonic Crystal structure of 5x5 structure plotted vs. the rods variation off the nominal position in the x-axis.

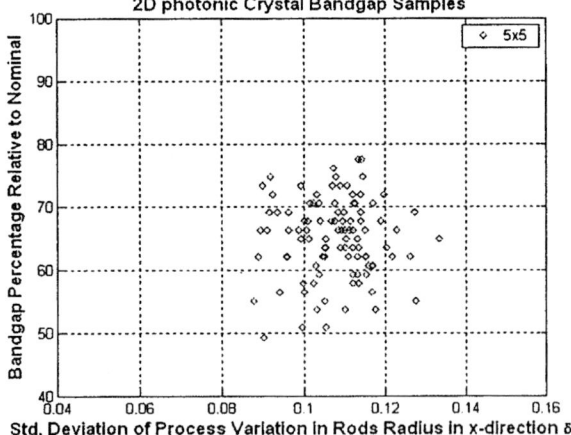

Fig. (14) Bandgap calculation samples of Square-lattice Photonic Crystal structure of 5x5 structure plotted vs. the rods radius variation in the x-axis.

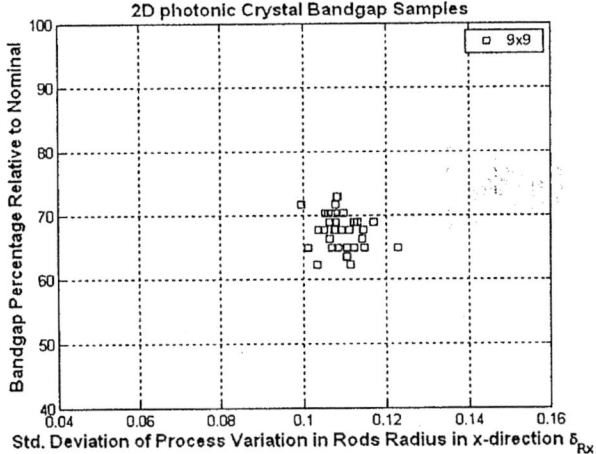

Fig. (15) Bandgap calculation samples of Square-lattice Photonic Crystal structure of 9x9 structure plotted vs. the rods radius variation in the x-axis.

From Figures (10-15) we can also conclude the following:

- As the standard deviation of the rods in the x-direction increases, the bandgap value decreases.
- The variation in rods radii has no clear trend on the bandgap value.
- As the size of the structure upon which the analysis is applied decreases, the scattering of the samples increases. (The 3x3 samples is more scattered than the 9x9 samples)

These data could be presented more quantitatively by using the Cumulative Density Function (CDF), as shown in Fig. (16).

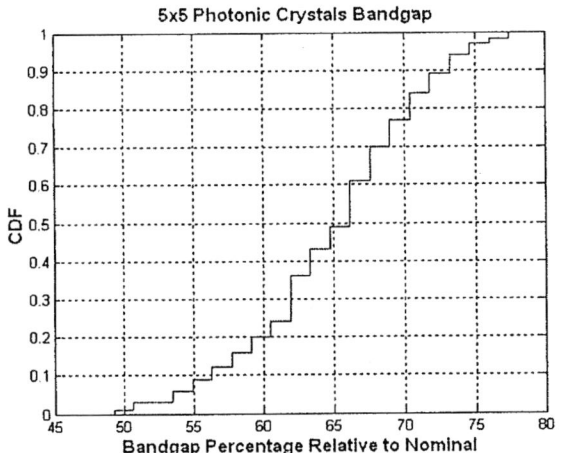

Fig. (16) The Cumulative Density Function (CDF) of the Bandgap of samples of the 5x5 Photonic Crystal Structure at 10% process tolerance

Actually, Fig. (16) can be used to calculate the yield of the fabrication process at a certain process tolerance. If we assume for example that the accepted bandgap variation is 60% of the nominal value, thus we can see that there are 20% of the samples having a band gap less than 60%, or we will have a yield of 80%. Another method to indicate the process yield is by calculating the Process Capability CP. The Lower Process Capability (CPL) of the photonic crystals bandgap variation process measures how much is the process meeting the lower specs limit (i.e. if the accepted bandgap value is 80% then L = lower limit = 80%). The bandgap variation is a single sided process; i.e. it has a defined lower limit and no defined upper limit because

the process variation causes a decrease in the bandgap value. The CPU of such single-sided process is equivalent to the CP of a double-sided process. It can be defined as:

$$ CPL = \frac{\mu - L}{3\sigma} \qquad (30) $$

where μ is the bandgap samples mean value and σ is its standard deviation.

For the 5x5 photonic crystal structure, for 10% process tolerance, we can find that the mean value for the calculated bandgap samples is 64.9014% and the standard deviation is 6.2080%. Then the CPL will be 0.263.

VI. CONCLUSION

In this work, we developed a Monte Carlo technique for the simulation of the effect of the fabrication process variations on the performance of the photonic crystal. This technique enables to evaluate the process yield for a given performance specifications. The technique is applied for the calculation of the bandgap of a two dimensional square-lattice photonic crystal. It is also used to evaluate both the CPU and the yield of the structure, taking into account its real manufacturing conditions of the crystal. This enables to optimize the photonic crystal design, using different geometries, e.g. honeycomb lattice, for a given required performance to enable the best design for manufacturability. It can also be extended to the performance evaluation of more complex photonic crystal structures, e.g. waveguides, filters and multiplexers.

VII. REFERENCES

[1]   M. Imada, S. Noda, "Recent progress of semiconductor photonic crystals", IEEE-NANO 2002. 26-28 Aug. 2002 Page(s): 217 - 218
[2]   M. Javanmard, R. Siemann, B. Cowan, "Photonic crystal laser accelerator structures", Particle Accelerator Conference, 2003. Proceedings of the Volume 3, 12-16 May 2003 Page(s): 1855 - 1857 vol.3
[3]   J. O'Brien, "Photonic crystal devices", LEOS 2003. The 16th Annual Meeting of the IEEE Volume 2, 2003 Page(s): 573 – 574.
[4]   X. Letartre, J. Mouette, J.L. Leclercq, P. Rojo Romeo, C. Seassal, P. Viktorovitch, "Switching devices with spatial and spectral resolution combining photonic crystal and MOEMS structures", JLT, Volume 21, Issue 7, July 2003 Page(s): 1691 – 1699
[5]   T. Badreldin, T. Saad and D. Khalil, "Yield Analysis of Optical MEMS Assembly Process Using a Monte Carlo Simulation Technique", IEEE, JLT Vol.23 No.2 Feb. 2005
[6]   K. Sakoda, "Optical Properties of Photonic Crystals", Optical Sciences Series, Springer, Berlin, 2001

Preparation of Polyimide Nanofibers by Electrospinning

Mingyan Zhang, Zhaoli Wang, and Yujun Zhang

Abstract—Polyimide nanofibers were obtained by electrospinning a solution of poly(amic acid)(PAA), a precursor of polyimide(PI) prepared in the laboratory, in dimethylacetamide(DMAc) at 14kV, followed by thermal imidization. Mechanical property of the resulting nanofiber non-woven membranes was investigated by tensile tests. Chemical structure and surface morphology of the non-woven mats were characterized by Fourier transform infrared spectroscopy (FTIR) and Scanning electron microscopy (SEM) as well as atomic force microscope (AFM) respectively.

I. INTRODUCTION

Polyimides, a class of polymer with the group of
$$-\overset{\overset{\text{O}}{\|}}{\text{C}}-\overset{|}{\text{N}}-\overset{\overset{\text{O}}{\|}}{\text{C}}-$$
, have been widely used in the fields of aeronautics、nuclear power、micro-electronics and et al. due to their excellent thermal stability and high mechanical properties, along with their good chemical resistance and electrical properties[1,2].Outstanding radiation resistance and electrical properties make PI fibers become a more potential material in the environment of high temperature and radiation compared with the other polymeric fibers. Research on PI fibers with superior properties has been receiving unprecedented attention along with the improvement of synthesis and the technology of spinning as well as the increasing requirements of the progress of the society.

Electrospinning is a straightforward method of producing fibrous polymer mats with fiber diameters in the range of ca. 0.05 microns to several tens of μm[3]. Since the resulting non-woven fabrics are consisted of nanofibers with high surface-to-volume ratio which have a broad range of applications such as tissue engineering[4], sensor[5], protective clothing[6], filter[7], etc, electrospinning has attracted great attention in the past decade. The advantages of electrospinning such as simple equipment, short time for operating and small amount of applied solution, etc make electrospinning become the focus of the research on

---

Manuscript received July 31, 2006. This work was supported in part by Heilongjiang Education Foundation under Grant NO.EO227, Harbin Youngs Foundation Program under Grant NO.2004AFQXJ048 and Heilongjiang Foreigner Technology Program under Grant NO.1053HQ001.

Mingyan Zhang is with Harbin University of Science and Technology, No. 4 Linyuan Road, Harbin, Heilongjiang Province, People's Republic of China (corresponding author to provide phone: 86-451-86392586; fax: 86-451-86392586; e-mail: taotaozh@hrbust.edu.cn).

Zhaoli Wang, is with Harbin University of Science and Technology, No. 4 Linyuan Road, Harbin, Heilongjiang Province, People's Republic of China (e-mail: wzl@hrbust.edu.cn).

Yujun Zhang is with Harbin University of Science and Technology, No. 4 Linyuan Road, Harbin, Heilongjiang Province, People's Republic of China (e-mail: zhangyujun2003@163.com).

---

nanofibers and a technology employed to prepare a variety of polymeric nanofibers

Although the technology of electrospinning develops very fast, not until the year of 2003 was the first paper on the application of electrospinning to the preparation of PI nano-fibers reported by Changwoon Nah[8]. In addition, two other papers on PI nanofibers were reported by Chan Kim and G.S.Chung respectively[9,10]. However no paper was found to focus on mechanical property of the resulting non-woven fabrics.

In this paper, PI nanofibers non-woven mats were prepared by electrospinning whose chemical structure and surface morphology were examined. In addition, its mechanical property was reported and compared with that of cast film prepared with the same solution and under the same condition of imidization.

II. EXPERIMENTAL

A. Materials

Pyromellitic dianhydride (PMDA) and 4, 4 -oxydianiline (ODA) were industrial products, purchased commercially from Insulated Materials Corporation of Mudanjiang, China and used without further purification. N, N-dimethylacet-amide (DMAc) was chemical pure reagent, purchased commercially from the same company and used with further distillation purification.PAA solution and PI cast film used in this study was synthesized by a well-developed procedure described by Mittal KL [11] and Mingyan Zhang [12].

B. Electrospinning

The experimental setup used for electrospinning process of this study is schematically shown in Fig1, which consists of an adjustable DC power supply, a syringe pump and a grounded target wheel wrapped with aluminum foil. For the process of electrospinning, polymer solution with varied PAA content was placed into a 10ml plastic syringe fitted to a stainless steel needle. Electrospinning voltage was applied to the solution by clipping an electrode onto the needle from the power supply. The flow rate of the solution was maintained by the syringe pump so that a pendent drop remained there all the time during electrospining.

C. Imidization

A serial of sequential thermal treatments of 1h (each) at 80 and 120℃,subsequently 30min (each) at 160, 200, 250℃,and eventually 1h at 300℃, was accom-plished for the conversion of the PAA fibers into PI ones

---

1-4244-0899-7/06/$20.00 ©2006 IEEE.

D. Characterization and Analysis

Morphology of the Electrospun fibers was observed by field emission scanning microscopy (FESEM; Phillip FEI Sirion) and atomic force micro-scopy(AFM, Nanoscope(r)—IIIa).Before mounted onto SEM plates, samples were sputter coated with gold. Mechanical properties evaluation was performed on a WDW electronic omnipotent testing machine produced by Changchun New Technology Corporation affiliated to Chinese Academy of Science with an extension rate rate of 50 mm/min to failure and specimen size of 70×10mm with effective length of 40mm.

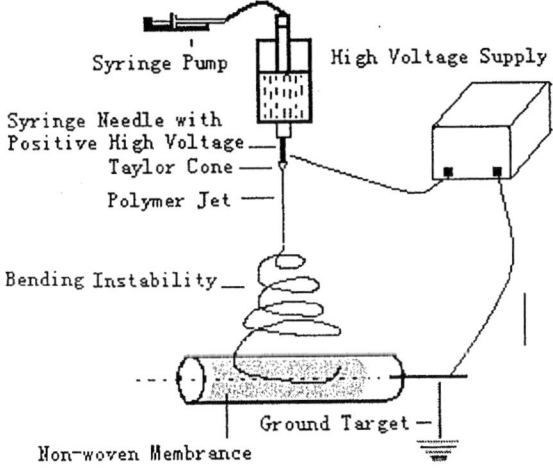

Figure 1 Schematic Electrospinning Set-up for this study

III. RESULTS AND DISCUSSION

A. Chemical structure confirmation by FTIR spectra

Molecular confirmation of the non-woven fabrics of PAA and PI was drawn from FTIR analysis. FTIR spectras of electrospun PAA and PI fibers were shown in Figure2.In the case of PAA, the characteristic peaks of C=O in an CONH group and the C-NH stretch band, can be seen at $1660cm^{-1}$ and $1550cm^{-1}$.When PAA is imidized, while the charac-teristic absorption of PAA at $1650cm^{-1}$ completely dis-appeared, the characteristic absorption peaks of imido groups at $1778cm^{-1}$(C=O symmetric stretching),$1727cm^{-1}$ (C=O asymmetric stretching), and $1379cm^{-1}$(C-N stretching) are evident, revealing a perfect imidization.

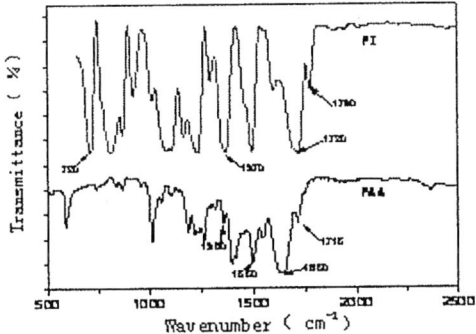

Figure 2 FTIR spectra of PAA and PI electrunfibres

B. Morphology of the non-woven fabric

1) Scanning Electronics Microscopy

Typically representative SEM photographs of all the non-woven fabrics for both PAA and PI are shown in Figure 3. As can be seen from these SEM photographs, the resulting ultra fine fibers had a diameter distributing ununiformly from tens of nanometers to about several hundred nanometers but generally below 300nm. As expected the diameter of the fibers increase with concentration, but all with significant beads for the proper reason being with the process parameters like the voltage, separation distance and inner diameter of the syringe needle. What can also be seen are many pores with different size existing indicating loose structure of the non-woven fabrics. From the two photographs, it can be seen that a certain extent of fiber orientation exists perpen-dicularly and oblique horizontally respectively.

a) PAA nanofiber non-woven fabric

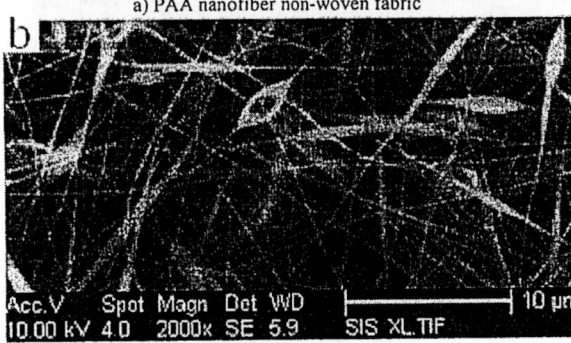

b) PI nanofiber non-woven fabric

Figure 3 SEM photographs of both PAA and PI

2) Atomic Force Microscopy

Figure 4 shows some of the typical AFM photographs for the non-woven fabrics of PI. From these AFM photographs, the same results as that of the SEM photo-graphs can be obtained.

The 2006 International Conference on MEMS, NANO and Smart Systems

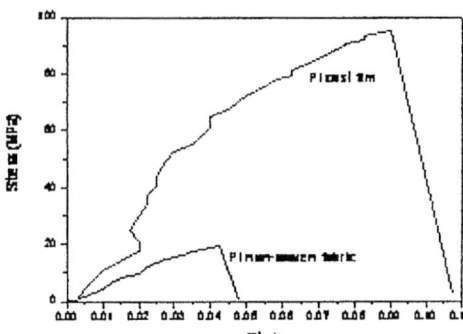

Figure 5 Tensile stress-strain curve of PI non-woven fabrics and PI cast film

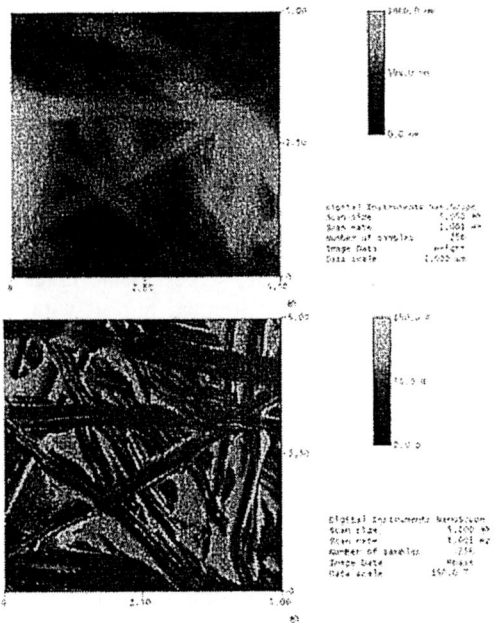

Figure 4 AFM photographs for PI

Both Figures 3 and Figure 4 indicate that the resulting non-woven fabrics are with micro-pores which makes them good candidates for various commercial applications, especially in the heat-resistant separation technology[12].

C. Mechanical Property

The tensile stress-strain curves of the electrospun PI nano-fiber membranes with 10wt% PAA and cast PI film are shown in Figure 5. From this figure, one can clearly see an evident phenomenon that the fracture strength of PI cast film is 94.6 MPa 4.8 times of that of PI non-woven fabric with 19.8 MPa, indicating that the mechanical strength of PI non-woven fabric is much poorer. This may be attributed to the fact that the PI non-woven fabric consists of fibers with significant beads and evident pores, which might have considerably reduced the cohesive force between the fibers of the non-woven fabric, and hence a poorer mechanical strength of PI non-woven fabric was obtained. From Figure 5, it can also be seen that the elongation at break of PI cast film is 8.95%, 2.1 times 4.24% of PI non-woven fabric. The smaller elongation at break might be a result of finer size of the fibers and molecule orientation existing in the non-woven fabric.

IV. CONCLUSION

PI nanofiber non-woven fabrics were successfully prepared by electrospininng and the obtained mats, which are consisted of nanofibers with significant beads, exhibit loose structure and a certain extent of orientation, which resulted its lower fracture strength and smaller elongation at break.

REFERENCES

[1] Wolfe JF. In: Mark HF, Bikales NM, Overberger CG, Menges G, editors. Encyclopedia of polymer science and engineering, vol.11. New York:Wiley-Interscience,1985.p.601-35.
[2] Sroog CE. Macromol Rev 1976;11:161.
[3] Larrondo L, Manley RS.J Polym Sci 1981;19:921.
[4] Matthews JA, Wnek GE, Simpson DG, Bowlin GL. Biomacromolecules 2002;3:232-8
[5] Wang XY, Drew C, Lee SH, Senecal KJ, Kumar J, Samuelson LA. Nanoletters 2002;2:1273-5
[6] Schreuder-Gibson HL, Gibson P, Senecal K, Sennett M, Walker J, Yeomans W, et al. J Adv Mater 2002;34:44-55.
[7] Tsaia PP, Schreuder-Gibson H, Gibson P. J Electrostat 2002;54:333-41.
[8] Changwoon Nah, Sang Hyub Han, Myong-Hoon Lee. Characteristics of Polyimide Ultra-fine Fibers Prepared Through Electrospinning. Polymer International, 2003, 52(3): 429~432.
[9] Chan Kim, Wan-Jin Lee, Kap-Seung Yang. Supercapacitor Performances of Activated Carbon Fiber Webs Prepared by Electrospinning of PMDA-ODA Poly (amic acid) Solutions. Electrochimica Acta, 2004, 50: 883~887.
[10] G S Chung, S M Jo, B C Kim. Properties of Carbon Nanofibers Prepared from Electrospun Polyimide. Journal of Applied Polymer Science, 2005, 97: 165~170.
[11] Mittal KL (Ed), Polyimides: synthesis characterization and application, Vols I and II, Plenum Press, New York (1984).
[12] Mingyan Zhang, Fang Wang, et al. Effects of inorganic components on electric properties of the hybrid polyimide film. Materials Science & Technology, 2004, 12(6): 576~578.

IEEE Catalog Number: 06EX1659
ISBN: 1-4244-0899-7

9781424408993